Disenchantment with Market Economics

European Anthropology in Translation
Published in Association with the Society for the Anthropology of
Europe (AAA)
Editor: **Susan Mazur-Stommen**, Research Anthropologist, University
of California, Riverside

This important new series introduces research from a new generation of
scholars working as Europeans and producing ethnographies of Europe. All
these works were originally published in their native languages and provide
an opportunity for fresh voices to be heard.

Volume 1
*Disenchantment with Market Economics: East Germans
and Western Capitalism*
Birgit Müller

To my daughter Nora, born in 1989

Disenchantment with Market Economics

East Germans and Western Capitalism

Birgit Müller

Berghahn Books
New York • Oxford

First published in 2007 by
Berghahn Books
www.berghahnbooks.com

Originally published as *Die Entzauberung der Marktwirtschaft.*
Ethnologische Erkundungen in Ostdeutschen Betrieben
by Campus in 2002.

Library of Congress Cataloging-in-Publication Data
Müller, Birgit.
  Disenchantment with market economics : East Germans and
western capitalism / Birgit Müller.
    p. cm.
  ISBN-10: 1-84545-217-8 (hbk.) -- ISBN-13: 978-1-84545-217-9 (hbk.)
  1. Manufacturing industries--Germany--Berlin--Case studies.
  2. Berlin (Germany)--Manufactures--Employees--Case studies.
  3. Germany--History--Unification, 1990.   I. Title.
  HD9733.8.B47M85 2006
  338.4'767094315--dc22
                                                              2006027976

British Library Cataloguing in Publication Data
A catalogue record for this book is available from the British Library

ISBN: 978-1-84545-217-9 (hbk.) 978-1-8454-506-4 (pbk)

Contents

Preface

In contrast to the enthusiastic official discourse of the Kohl government, I had a sceptical attitude to the rapid political unification of both German states. Educated by teachers from the 1968 generation, I lacked a feeling of national pride or even a feeling of German national identity. Although I had lived in West Berlin from 1982 I felt closer to France (where I had studied) or to Austria (where I spent part of my childhood) than to the GDR (German Democratic Republic). The wall that surrounded West Berlin, the strict border controls and visa requirements were effective in dissuading me from visiting the beautiful surroundings of Berlin. I started to discover East Germany only at the end of the 1980s when contacts with East Germans became easier. I became interested in the efforts of the newly awakening opposition movement to develop alternatives to the GDR regime. With many West Berlin activists and intellectuals I shared in 1989 the hope that a discussion about social alternatives could emerge that would bring new momentum to social movements in West Germany. When real existing socialism collapsed and the opposition in East Germany declared that it did not want to fall immediately into the wide-open arms of capitalism, expectations were high. However, this hope was rapidly thwarted by what Habermas called the 'chubby-cheeked DM nationalism' (1990: 205).

Ever since my research in West Berlin collective enterprises in the first half of the 1980s I had wanted to do research in a 'real' people's owned enterprise. When the Wall fell, I was curious to find out how working people in the GDR would experience the market economy and what ideas they would approach it with. When I finally fulfilled this wish in the spring of 1990, people-owned enterprises had almost ceased to exist. However, memories of the planned economy were fresh and the people working there were keen on sharing them with me. I began my explorations together with the students from the research seminar 'Transformations of the Culture of the Everyday in East Germany' that

I organised from April 1990 to July 1991 at the Institute of Ethnology at the Free University of Berlin. Rainer Bachmann, Gregor Noack, Hildegard Weber, Dagmar Heymann, Christina Katzer, Britta Heinrich, Thomas Hammer and Katrin Arndt took part in the first heated debates. I shared with Katrin Arndt the highs and lows of doing fieldwork at Taghell[1] and the interviews and fieldnotes.

I would like to thank all employees at Taghell, Stanex and Hochinauf who shared with me their good and sometimes disagreeable memories and had long talks with me about their appraisal of the present and their expectations for the future. I often stood in their way when they tried to work and they allowed me nevertheless to look over their shoulders. My special thanks are for Mr Stolz, who did not hesitate, in sharing his unvarnished opinion with me.

The German research foundation DFG provided me with a generous research grant that made it possible, not only to carry out research in Berlin, but also to continue it later in Moscow. I started writing the book at the Centre Marc Bloch in Berlin and continued it at the CeFReS in Prague. Discussions with my colleagues of the LAIOS in Paris, Marc Abélès, Catherine Neveu, Irène Bellier, Jean-François Gossiaux und Pierre Bouvier helped me digest the material collected.

Many discussions with colleagues and friends have encouraged and helped me with this manuscript. I wish to express my sincere gratitude to Emmanuel Terray, Simon Clarke, Claire Wallace, Michal Bodemann, Miklos Hadas, Michel Pialoux, Sighard Neckel, Helmuth Berking, Richard Rottenburg, Ivana Mazalkova, Ed Clark, Anna Soulsby, Pavel Romanov, Veronika Kabalina, Albert Hirschman, László Bruszt, Effi and Ewald Böhlke. Long conversations with Wolf Schäfer helped me to understand everyday life in GDR enterprises. Horst Froberg contributed his poems and his ironic view on things. Discussions with my father helped me to understand the inner life of a multinational corporation. Claus Offe went through the entire manuscript and sent me suggestions I could act on. My friend Henk Raijer did as much as anyone to see the German version of the manuscript through to completion.

The English version was produced in cooperation with John Bellamy, who delved deep into the vocabulary of the planned economy to translate Chapters 1, 2, 3, 4, 6 and 10, as well as the introductions and conclusion; Jennie Challender, who translated Chapter 5, and Kathleen Repper who did an excellent job on Chapters 7 and 8. My special gratitude is to Maeve Oloham of the University of Manchester, who put me in contact with and trained my team of translators.

I wish to thank my partner Terry Boehm for the keen interest he took in the planned economy while helping me revise the English translation. It is a particular pleasure for me that my daughter Nora, born in 1989, started to read the book when she chose as a school research project 'memories of real existing socialism'.

Notes

1. All names of persons and enterprises in this book are pseudonyms.

Introduction

When the Berlin Wall fell in the autumn of 1989, the competition among political and economic systems, which had determined the course of the twentieth century, seemed decided in favour of the market economy. The end of central planning, following decades of stagnation, generated new enthusiasm. It produced hope and optimism, inspiration and excitement. To workers and employees in the companies of East Berlin it conveyed a feeling of unlimited possibilities, of joy in action, and in the regained meaning of life, awakening utopian visions for the near future. Once the power of the regime seemed to have withered away, workers openly criticized their managers, blamed them for errors in administration, for their incompetence and political dogmatism. They openly voiced criticism that had long been circulating covertly, and they made plans for how everything would be better and different in the enterprise.

What, however, were working people able and willing to change in the enterprises following 1989? How did they reflect upon their actions and upon the possibilities that now opened up to them with the collapse of the old authoritarian social structure? I have tackled this question since 1990 by looking at three enterprises in East Berlin. In this book I attempt to unveil some of the myths and promises of the free market economy, by looking at them from the viewpoint of those who experienced the collapse of the planned economy in their day-to-day working lives. At the centre of this book stand the daily lives and the personal experiences of those working amidst the confusion of the political and economic upheavals. The central argument engages the issues of power and personal autonomy in the enterprise before and after the *Wende* (turnaround). The political and ideological mechanisms of influence and power in the planned economy are contrasted with the ways in which those mechanisms work in enterprises of the market economy. I will demonstrate by which informal and institutional means the management and the workforce renegotiated their relationships, and in so doing how they redefined concepts of solidarity, responsibility and public welfare. The brief historical moment of the German Wende from autumn 1989 to summer 1990 became for the workers a crucial experience that called into question ideas about the rationality of the market economy, the legitimacy of ownership, and democratic control. Their ideas of 'market', 'private property' and 'democracy' became objects of my research, and are questioned as to their ideological nature.

The book consists of three parts. The first part is dedicated to the 'memories of the planned economy'. Power relations conveyed by the plan, socialist competition and the party are reconstructed in this part of the book from life stories, memories and written material from the time of the planned economy. The second part analyses the strategies of the individual in the enterprise following the collapse of the macrostructures of the planned economy. New everyday practices developed during this period of collapse. However, they have little in common with a rational economic environment, which was supposed to take shape with the market economy in East Germany. The third part analyzes how the business strategies and management philosophies of a multinational firm were received in a former 'people-owned enterprise' (*Volkseigener Betrieb*). The book finishes with a comparison of the mechanisms of ideological power and material control in the socialist planned economy and the globalized market economy.

The institutional transformations taking place within the enterprises make power relations visible, which in the everyday running of the enterprise are generally embedded in routine and habits (Clegg 1979). Staff in the East Berlin enterprises were able to perceive the power relations in the companies of the market economy differently and more keenly than their colleagues in the West who already lived with such relationships in their everyday lives. Since the old structures of control no longer appeared natural and the new ones were not yet anchored in routine behaviour, these power relations could be clearly observed (Clegg 1979: 147). This study provides an insight into the micromechanisms of power (Foucault 1975, 1977, 1986, 1987) and how they are interpreted by the workers themselves (Scott 1990; Herzfeld 1997). In previous studies of the Wende in Eastern Europe 'large' institutional changes have been considered, rather than these micromechanisms of power, although they help to explain these changes and both of them are mutually dependent.

In order to analyse how forms of power and control differ in socialist enterprises and in the market economy, Foucault's concept of power is useful. Foucault does not perceive power as a thing, possession or chance, but as a resource for everyone in his or her sphere of action:

> Power comes from below; that is, there is no binary and all-encompassing opposition between rulers and ruled at the root of power relations, and serving as a general matrix – no such duality extending from the top down and reaching on more and more limited groups to the very depths of the social body. One must suppose rather that the manifold relationships of force that take shape and come into play in the machinery of production, in families, limited groups, and institutions, are the basis for wide-ranging effects of cleavage that run through the social body as a whole. (Foucault 1977: 115)

Foucault explains historically how direct violent forms of exerting power in sovereign societies were discarded in disciplinary societies, where social discipline was implemented through other means, such as physical drill and total surveillance (Foucault 1986: 241). Modern industrial societies are disciplinary societies in which social discipline is internalized and accepted by the social participants as the norm. Therefore it no longer has to be imposed from outside. The relationships of control in the enterprises of the market economy emphasize individual self-discipline far more than would have been the case in socialist enterprises. This brings us to the central issue of this book: in the transition to the market economy, has the workforce gained or lost in personal autonomy and self-determination? Has the market economy brought them the freedom they had hoped for?

The exertion of control is always a mutual process between those who exert it and those who tolerate it and more or less demand it. To show this relationship, I examined to what extent the workforce in East German enterprises attempted to change the mechanisms of control and discipline within the enterprise and to influence economic decisions. Nonconformist discourse about power and justice in the workplace already existed in the planned economy. The hidden transcripts (Scott 1990), which were circulated during the time of the GDR, took on the form of jokes and anecdotes about poor socialist planning and inefficiency. There were also rumours about corruption and embezzlement, all of which would be whispered within intimate circles of confidants and colleagues. At the time of rapid social change they came to light as collective interpretations of social reality (Scott 1990: 204ff.). Above all, they were directed against the directors of enterprises, *Kombinate*[1], and against representatives of the party hierarchy. Shortly after the fall of the Berlin Wall they were translated into actions that were 'difficult to categorize and even more difficult to institutionalize' (Engler 1995: 76).

The basis for my analysis of the micromechanisms of power is ethnological field studies undertaken in three East Berlin enterprises: a lamp factory, an enterprise building roundtable assemblers and an enterprise constructing lifts. The initial period following the fall of the Wall (1990–91) is predominantly documented in the first two, and it was among the workers there that the debate about the planned economy, which had just recently become part of the 'past', was at its liveliest. The third enterprise was purchased by a multinational corporation that also owned a branch in West Berlin. In all three enterprises, I spent most of my time on the shop floor, observed the interaction there and had prolonged conversations both during and after working hours. In the four years from 1990 to 1994, I pursued the dramas about power, ownership, and worldviews that unfolded and fed on past occurrences, present concerns and ambitions for the future.

The radical changes following November 1989 affected all levels and dimensions of life in socialist enterprises: the production and distribution of goods, social support for the employees and for their families, party political events, socialist competition, social organisations, work routine, holidays, and the individual psyche. Without making the enterprise into the object of research, I investigated *within* the enterprise how constellations of power were dissolved and reconstituted, how preconceptions of order and social justice were turned upside down, how the use of time and space changed, as well as the ways of coping with machines, products and colleagues in everyday life. The usual and unusual events, along with employees' comments explaining these to me, made it possible to observe their individual and collective strategies and to become familiar with their perspectives, dreams and plans. I was interested in the extent to which they could bring influence to bear on the fundamental transformation that took place in their enterprises, either by supporting, tolerating, or fighting it. To what degree did the people in the enterprises after 1989 shape the social and political structures?

The political Wende in the GDR that surprised observers and participants in 1989 happened at a time when the Western world had also entered a phase of rapid social and economic transformation. From the 1980s Western industrial nations were being shaped by the consequences of a new technological revolution occurring primarily in the area of communications. Information can be sent right round the globe at the speed of light through optic fibres, and the smallest electronic elements now enable worldwide networking of production and planning processes together with decentralised control of them.

At a time when decentralised planning and the complex networking of groups and individuals were being made technologically possible, GDR society collapsed under the burden of its centralist and de facto arbitrary administration by elite party members. Technologically, especially in communications and computers, the GDR had missed out on the decisive years of the 1980s. The East German economy had to prepare itself after 1989 for a quantum leap out of a closed system into the globalisation of relationships both in production and in the market. Overnight its products had to rank among others in worldwide competition. Not only was production in East Germany no longer secured by the socialist state but the protection through borders, customs duties, exchange rates and subsidies also became increasingly fragile in the 1990s due to international treaties on free trade and the lifting of currency barriers in the European Community.

With the disappearance of central state control, the security of a closed and narrow view of the world disintegrated too. From an age of duplication using matrices and carbon paper, and of direct face-to-face communication, as was usual in a GDR enterprise prior to 1989 – particularly at a time when it was better if the state did not overhear what was being said – East German society entered into the age of the internet. Continuous communication is now not merely desirable, but demanded. Communication and mobility, both conveyed through the omnipresent medium of money, turned into a 'must-have' of the new society.

The planned economy had been accompanied by a code of conduct, which was tied to the moral system of values in socialism. This code was not identical to the publicly disseminated ideology; it even contradicted it in many respects. Whereas the socialist ideology demanded competition and achievement, it was in fact a levelling out of performance that gained acceptance in enterprise practice. In spite of collective and individual bonuses, which were often granted not only for performance at work but also for good political conduct, the differences in income between the managing director and the average worker were small. Money, too, had little appeal because of the scarcity of goods that could be bought.

Along with the end of the planned economy came a change in the function and meaning of money for social and economic relationships. Social domains, which could be regulated by money, expanded, rendering the close-knit personal networks for exchange and mutual services unnecessary. Members of society became dependent on money for solving numerous everyday problems. High income promised high social status. Social distinctions emerged that had not existed in the planned economy of the GDR.[2] The social code of conduct that demanded individuals keep themselves in check, remain unnoticed and not stand out from the collective, abruptly lost its basis after the fall of the Wall. With the collapse of the planned economy a process of selection began that quickly threw unproductive, inflexible and passive employees onto the scrapheap. Being noticed and standing out positively from the masses now became essential.

Whereas in the GDR the production of goods for society was held in high moral esteem and the enterprise was regarded as the 'nucleus' of socialist life, the new demands of the market economy no longer appealed to a social mission. Successful competition in the marketplace turned into an aim in itself. Products were manufactured worldwide in those locations that were cheapest, where wages and social benefits were low and the organisation of trade unions weak. Companies in industrial nations of the West demanded that their workforce be willing to adapt to the conditions of competition, accept closures and redundancies and at the same time develop creativity and quality awareness.

The consequences of rapid institutional reform and of changing the legal framework for all economic and social activities exceeded any expectations that employees could have had before 1989. Were they ready then to use their own ideas and initiatives to flesh out the institutional framework, which the laws and institutions of the Federal Republic of Germany put forward? Or did the institutional establishment of power and ownership stifle all the approaches and initiatives that blossomed shortly after the fall of the Wall?

From 1 July 1990 the West German Deutschmark became the medium of exchange for enterprises in East Berlin. During the privatisation of the people-owned enterprises through the 'trust fund', *Treuhandanstalt*, the decision makers of the planned economy negotiated with the economic and political decision makers of West Germany. This left little leeway for experiments and for the hazy preconceptions employees had had of the changes that should take place in their society. However, East German society did not only change from the top down. It was everyday action both inside and outside the enterprise, often without any intention of reform, which shaped and changed East German society. West German institutions had to find a certain degree of recognition, acceptance, or even support among East Germans and be in keeping with their norms and values in order to be able to influence society in the long term. There were tensions present in day-to-day life between values and norms that had arisen, on the one hand, from everyday life in the planned economy and, on the other hand, from the expectations about the market economy. These tensions also included those values and norms that the West German reformers anticipated. They often found expression indirectly through endless comparisons contrasting characteristics and behaviour attributed to *Wessis* and *Ossis*.[3] Attitudes also developed, changed and became differentiated in the transformation process, not least because of the individual success or failure people experienced in the new social system.

After unification, in conversations on the shop floor and in offices, the employees constantly analysed and commented on the consequences of and conditions behind managerial decisions, trying to understand the reasons for their actions and for their nonaction. Yet, the results of these actions did not necessarily correspond to what was originally intended (Giddens 1987: 61). Such undesired consequences of everyday decisions made at the time of the Wende are a focal point of this book.

Since West German institutions regulated the process of transformation in East Germany, any undesired consequences of this transformation could in everyday discourse be blamed on the 'West Germans'. This, however, concealed the fact that the directors and staff actually made the West German institutions their own and did not merely put up with them passively. Who were the people who welcomed the West German institutions and profited from them, and who were the ones to be surprised and overwhelmed by them? Did those who had wished for the West German institutions to take over actually manage to profit economically? Why were some workers in the GDR in favour of institutional innovation, while others resisted it? Which experiences and at least semiofficial rights of status did they appeal to? How did they link these to the experiences and perceptions of the new enterprise regime?

The starting point of my analysis of power relations was the sentiment many employees shared that since the demise of the GDR they had become more solitary and less capable of collective action. For the people I talked to, this came as all the more surprising as in their eyes the legal situation had changed considerably: in the Federal Republic of Germany they now had free choice of workforce representatives and the right to strike, neither of which were available to them under socialism. The right of workers' participation in privatised enterprises was laid down in the legislation of the Federal Republic. Nevertheless, the employees rarely made use of this. They were unaware of their rights and mistrusted the trade union leaders, who led the organisations that succeeded the old socialist trade unions. In the privatised enterprises they initially recreated the structures of communication that they had been familiar with in the planned economy.

Although in the first period of radical social change after the fall of the Wall, the rigidity of social structures seemed to relax, it then hardened again in the years that followed. The very first critics in the period immediately after unification demanded changes and brought their managers to account, but it was mostly those who formerly wielded power who now exploited the opportunities of the new system to their advantage and bolstered their positions of power. Since the power that the directors exerted over their subordinates at the time of the planned economy was of a political rather than economic nature, how were they able to maintain their hold after the party structure in the enterprise had been dissolved? What motivated these directors and how did they gain acceptance and legitimise their actions towards a workforce to whom they had preached Marxism-Leninism for years on end?

The complex mechanisms of cultural adaptation and refusal in the enterprises differed depending on whether their former director privatised the enterprise or whether West German firms or multinational corporations purchased it. In the latter case, a 'corporate philosophy' was offered to the employees as a model for identification, which was to explain and legitimize the new power relations and principal values. How the mechanisms of 'corporate philosophy' compare with those of 'socialist competition' exposes the link between work and ideology in both political systems.

Methods of Field Research

The perspective of the anthropologist arises from the tension between proximity and distance, from the close interaction with the people one is working with, and from the essential requirement to maintain the capacity for reflection and self-reflection. The rationale of the research and the methods employed are mutually dependent. In this way, social anthropology faces the challenge of absorbing the subjectivity of the researcher into the analysis. At the same time, the research has to be methodologically transparent. When analysing their own society social anthropologists face an even greater challenge since the link between their own ideas and the subject is all the more complex and the social anthropological self-analysis of the researcher is even harder to achieve because the individuals who are the focus of interest are neither necessarily different, nor alien.

In this respect, the social anthropological research in East Berlin had many traps set for me. Upon beginning my research, I saw the employees of the enterprises as 'others'. In the summer of 1990, when the structures of the people-owned enterprises were still predominant, they were unfamiliar to me as were the employees, who seemed eager to explain the planned economy and the special patterns of behaviour that went with it. They virtually insisted that a culture of its own had developed in the GDR, one that the West Germans would have to understand and take into consideration. All the same, this seemingly clear distinction between East and West became less evident as the research progressed. The construction of the other as a subject of research stood in direct contradiction to the process of the unification that was taking place during the same period. East German society received the institutional framework from the West Germans. They both adapted to it and adapted it to suit them. The 'others' lost their separate social context – the context in which I would have been able to enter, as a well-meaning social anthropologist, in order to understand and get to know them. They developed their own dynamic for dealing with the past and changed their rhythms of work and life, along with their cultural references. West German society changed too – this occurred in West Berlin considerably more rapidly and comprehensively than in the rest of West Germany. The points of contact, where the two societies met, shifted constantly. The enterprise was the point of distillation of these social developments, which were unfolding rapidly.

I had begun my research with simple categories, but these soon disintegrated. An example is the category of the 'East Berlin worker'. In 1990 the category 'worker' still retained strong ideological and political connotations in East Germany. It continued to be burdened by the memories of the 'workers' and 'peasant state'. Then, only one year later, this had almost completely disappeared. A whole new category crept in: 'the unemployed'. People who, in 1989, were certain of remaining workers all their lives, became traders, representatives, self-employed craftsmen and taxi drivers. 'East Berlin workers' who after 1989 had begun to work in the West Berlin firm Hochinauf, returned to East Berlin as 'West Berliners' with the merging of West German and East German production. During the course of this long-term study, not only did my relationship with the workers in the enterprises evolve, but I was also affected by fundamental political and social changes, which occasionally placed us, as West Germans and East Germans, in different political camps.

Thus, the results of the research do not attest to 'neutral' objective findings, but rather to the process of how these changes came to be interpreted. Through observing workers in action, through being constantly present even in moments of tension and conflict, participating in birthday parties, work conferences and staff meetings, I could see how their stories and explanations related to their actions. At the same time, however, I became a sounding board for the workers' self-portrayals, which enabled them to position themselves in this new reality.

There is much discussion in social anthropological literature about the first time contact is made in the field, and how the people who make this access possible have an impact upon the development of the research. I chose a path through the hierarchies that was typical for the old GDR – from the Ministry of Light Mechanical Engineering downwards – in order to gain access to the people-owned enterprises. The most difficult part was getting past the barriers and checkpoints at the entrance to the ministry at the beginning of May 1990. Once the barriers had been passed, it turned out that the ministry had already ceased to function. The employees still sat in their rooms – but they no longer worked. I had penetrated into the control centre of the GDR economy, but its power had come to an end. The employees of the ministry had no need to guard any secrets from me. They made time for me, looked up addresses and telephone numbers of enterprises they believed were successful enough to survive in the market economy, and recommended me to friends who were enterprise directors there.

The fact that I had been sent from the ministry ensured the doors of the first two enterprises, Stanex and Taghell, were wide open to me. The directors permitted me to move freely around the enterprise and to talk to employees. It certainly was not any great interest in my work that motivated them to admit me into the enterprise, but as a social anthropologist, I was in their eyes nothing more than quite simply a sign of the times: a Westerner, whose intentions they could hardly fathom, but to whom they did not want to bar access because they wanted to avoid the impression of having something to hide.

When I visited the lamp factory Taghell and the manufacturer of assembly machinery Stanex for the first time, in July 1990, the workforce and supervisors met me with real openness. I was able to have long conversations and take notes. The conflicts, alliances and antagonisms in the enterprise were gradually brought to light. Although members of the workforce and the management each tried to convince me of their respective positions, they did not prevent me from hearing contrasting accounts. The supervisors of the people-owned enterprises did not know yet whether they would be able to hold on to their jobs once the enterprise had become a limited company. On the other hand, the workforce still questioned the political past and the present economic decisions of their managers, but they had by this time already given up the idea of taking over management themselves. This stalemate was essentially ideal for ethnological research: memories of the socialist past still retained enough conflict potential in the present, so that the managers and the workforce inundated me with reports and accounts about the planned economy, while at the same time the power structures from the past had become temporarily ineffective.

Research became increasingly difficult during 1991 when, in light of the growing economic difficulties, management began to cover up its decisions and company policy. The socialist directors had become managing directors and tried to become owners of the enterprise. They no longer wanted me as an observer in the enterprise. Access to the first two enterprises became harder, forcing me to interview the employees mainly outside of the premises. Therefore, in the summer of 1991 I extended the field to a third enterprise, Hochinauf, which had become part of a multinational company. A member of the board from West Germany had acted as an intermediary to put me in touch with the senior management in West Berlin.

The issues that arose from observations in the workplace and from the interviews moved beyond the realm of the enterprise. The trivial matters, the everyday changes, the banal commentaries and observations, which were often only made offhand, all reflected the greater social context. These daily occurrences and accounts can be subjected to a variety of perspectives (Rosenthal 1996: 142). Social interactions such as conflicts suggest different meanings depending on whether I consider them in a 'microscopic' sense in all their details, or whether I regard them from a broader 'macroscopic' perspective. To provide just one example: the difference in pay between the East Berlin workers and West Berlin workers in the Hochinauf factory located in East Berlin can be considered from a wider 'macroscopic' perspective as a consequence of the unification treaty that temporarily established in East Germany an economic zone with a lower pay scale. Viewed from the 'microscopic' perspective of the workshop, it appears to be an attempt by the West German management to discriminate against East Berlin workers and became the catalyst for normative discourses about productivity and justice.

Official statistics, legal documents and reports provide valuable background information and without these a 'macroscopic' perspective cannot be adopted. This information taken by itself makes management decisions seem impersonal and almost objective. For this reason I made an effort to outline the complexity of the decisions, including those made higher up in the company, and to portray the decision makers as complex individuals. With this in mind I did not carry out interviews only within the company, but also with the management of the concern, with the administration of the Treuhandanstalt and with members of the supervisory board. Although I concentrated my research on the enterprise and made most of my observations there, the factory was merely the location where I carried out my fieldwork and not the 'natural' boundary of the research domain. The relationships of the participants in the study and their interpretations pointed constantly beyond the enterprise: to their family lives, their domestic circumstances, their political commitments, and stories of their past.

In the phase of restructuring the enterprise, conflicts from the socialist past reemerged and the members of staff attempted to settle their personal and political accounts. Some events from the past were extensively commented upon and described; others were suppressed. It was often only when we were looking together at historical documents, like brigade diaries (*Brigadebücher*), records from the party organization of the enterprise (*Betriebsparteiorganisation*), minutes from the meetings of the enterprise trade union committee (*Betriebsgewerkschaftsleitungs-Sitzungen*), that painful memories were vividly recalled. Written documents triggered interpretations of the past from the perspective of the present, which often told just as much about recent experiences as about the past itself. After as little as one year, shifts could already be detected in the way the past was being interpreted.

Following several intensive phases of field research, during which I remained in the enterprise on a daily basis during working hours, I continued to visit them regularly: I arranged interviews and kept myself informed about the latest events. When the third enterprise, Hochinauf, was closed down by the multinational concern in 1996, I finished collecting material. The bankruptcy of the enterprises or the relocation of production was of course not the 'natural' conclusion to the field research, for the people continued to exist, either going on to find new jobs or becoming unemployed. Along with the closure of the enterprises went the disappearance of a location where I could continue to observe. To go on and accompany my subjects on their new paths in life would have provided material for another project.

Introducing the Enterprises and the
Participants of the Study

Stanex

Stanex developed in the 1960s from a research institute for cybernetics. In the years when the GDR economy was reformed into what came to be called the New Economic System,[4] cybernetic ideas of systemic self-regulation were applied to a new 'scientific organisation of production'. In 1968 the engineers of the institute began to transfer the principle of self-regulation as a form of crisis management to other enterprises, and from these ideas they went on to develop their own automated products. One of the initiators later became the director of the department of assembly automation of Stanex, Dr Schöpf, a staunch party member with a doctorate in planned economics.

I chose the department of assembly automation with one hundred employees as an area for research. It had its own administrative structure and functioned more or less independently within the enterprise Stanex, which had two thousand employees in all. The employees and the directorate of this department were highly motivated and convinced that they had a product of excellence. The debate about the political aims of the planned economy and the ideals of socialism formed part of their everyday conversations.

In the first few years, the research engineers were unable to bring any of the products they had developed into production. The mechanisms of the planned economy, which considered each innovation as a new complication, put a stop to their ambitions. It was not until the mid 1970s that Schöpf, thanks to good political connections, met with success in convincing his superiors in the enterprise, the Kombinat and the party to put their invention into production.

The invention was a roundtable assembler based on the principle of twelve interconnected assembly units, which had to be adapted to the requirements of each customer. The assembly units were vibrating tables that transported components for assembly to be fitted, screwed or pressed. Machines were constructed, for example, for putting together grounded electrical plugs or, in combination with a second machine, to fit pieces into a circuit board. Since every machine was custom designed, new problems arose each time that usually had to be solved on the spot. During the assembly process, the fitters and the designers worked closely together. Their relationship was an egalitarian and collegial one because they had to find pragmatic solutions together, instead of laying emphasis on hierarchies. At the time of the planned economy, the enterprise had successfully established a wide network of links with enterprises delivering special components that were crucial to the production of the roundtable assemblers. These machines were in great demand. The department director granted them above all to those enterprises that were prepared to deliver special parts in exchange for them.

When I began my research in the enterprise in May 1990, the department had already lost a large part of its business partners. The electric and electronic industries in the GDR were in crisis and assembly machinery was no longer required. New customers in the West German marketplace had not yet been won over.

Taghell

The small enterprise producing brass lamps in a Berlin suburb (let us call it Taghell) stood in contrast to Stanex in many respects. It produced highly desirable consumer goods for national consumption, as well as for export into other socialist and non-socialist countries. The workforce was mostly composed of non-party members. Accordingly, the level of both ideological indoctrination and political motivation of the employees was low. Most of them worked here because the job was well paid or because they lived nearby. Gossip about the directors asserted that they had come to the enterprise because there they could earn money with the lamps through unofficial channels.

Taghell had emerged from small family businesses that were producers and suppliers of lamps, which in the 1960s had gradually gone through the various phases of nationalisation. Through high taxation of profits (peaking at 80 percent) and lack of access to investment goods, the small firms became increasingly dependent on the state, acquired debt and had to surrender shares to the state. In 1972 it took over the enterprises completely in the final wave of nationalisation. Even after 1972, as state enterprises, the lamp producers continued to face grave difficulties in obtaining investment quotas to renew their machinery. It was not until 1978 that the situation of the enterprise suddenly began to improve. Although brass reserves in the GDR were low, the political decision was made to provide the enterprise with some of these resources for the production of their brass lamps.

These highly desired lamps were particularly suitable for making friends among state officials who were responsible for supplies. The District Economic Affairs Council was responsible for the enterprise up to 1984 and in return for receiving brass lamps for its members, it granted Taghell opportunities for investment. This changed in 1984 when the enterprise was incorporated into the Kombinat XYZ, where the combine management put any investments intended for Taghell aside so that they could be allocated to the main enterprise of the Kombinat, for which the director of the Kombinat was directly responsible.

However, according to GDR standards, the enterprise continued to remain extraordinarily successful even after 1984. From the mid 1980s on, a type of lamp was developed that met with approval in the West. The economic success of the enterprise – that is the fact, it not only manufactured popular consumer goods, but that it exported too – attracted executives from other, more important enterprises. When the new directors arrived, they frequently brought along a new team of their own, who would take up managerial positions (Arndt 1997). Particularly influential was the 'low-voltage gang' – who were given this name by their colleagues because they had all come from the people-owned enterprise VEB-Elektrik to Taghell, where in 1978 they occupied the positions of production director and technical director.

In 1989, ten small production units for lamps, components and lampshades, which all belonged to Taghell, were spread out over the city of East Berlin and its outskirts. The primary production building, where the administration was located and where I carried out my investigation, was situated in the centre of a Berlin suburb directly by the river. Along with its headquarters of 120 employees, Taghell owned a tool factory and a department constructing means of rationalization[5] with fifteen people; a development and prototype construction department with seven employees; a lamp shade factory with forty employees – mainly women; a warehouse storing finished products; a hand polishing plant employing three men, along with another one employing five men on the other side of Berlin; as well as a wood-turner's workshop that was run by a married couple. In 1986 two more enterprises were added to the people-owned enterprise, although they had nothing to do with producing lamps: a metal-pressing workshop with four subsidiaries, and a factory producing variable condensers with three subsidiaries. In 1989 Taghell was now composed of fifteen enterprise parts with a total of 340 workers. An employee hit the nail on the head when he called this accumulation of enterprises a *Hüttenkombinat* (a combine of shacks) because they all used primeval technology for production.

When Katrin Arndt and I visited the enterprise for the first time in the summer of 1990, brass debris and machine parts cluttered up the yard. The machine shop on the ground floor was untidy and the machines, standing in pools of oil, were making a deafening racket. With few exceptions, only women worked on the first floor of the rear building, where they were assembling the lamps. The room was bright and friendly, equipped with a generous kitchenette, and had a view onto the dirty yard and the river Spree below. Male machinists, paid on an hourly basis, fabricated the parts that were then passed on to the female assembly workers, who were all paid on a piecework basis. In response to their male colleagues in parts production, who had adorned their kitchenette with pin-ups of naked women, the female assembly workers had stuck a larger-than-life sized picture of a naked youth on the end wall of the assembly shop. On the second floor was a large canteen, which also served as an assembly room, and the floor above contained the surface treatment of the brass and wood parts. It was here that the brass fittings were polished by hand at small polishing grates, a task that required great skill and endurance. Afterwards they were varnished by less qualified assistants and made shinier with a nitro-diluted solution. Although the room containing the nitro-diluted solution had been fully renovated as recently as 1982, it remained the devil's workshop of the enterprise. Finally, they were polished again by other women and then packaged.

Hochinauf

During the GDR era, the lift firm VEB Lift was a showpiece project. It benefited from the apartment building programme of the 1970s, which paved the way for the expansion of lift production in the GDR. An investment programme was approved for VEBLift that enabled the construction of a main building and an assembly hall, along with a paint workshop, costing in total around 25.5 million GDR marks. Franz Oswald, who was in charge of this investment programme, had to doggedly convince the relevant political players to liberate the resources for this plan to proceed. Each year he predicted figures, which were overoptimistic and were never achieved, to convince them to continue investing.

Although VEBLift was a privileged enterprise and its main production line had political priority, it also produced a great many other products. Just like other GDR enterprises, it had to manufacture an obligatory contingent of consumer goods. The employees produced car trailers and hose trolleys for garden hoses. These were highly desired among the owners of holiday cottages (*Datschen*). In 1983 they began with the development of an escalator prototype. After it was completed in 1989, it was immediately scrapped. Another subsection of the enterprise producing trough chains for openpit lignite mines had already been stopped before the Wende.

VEBLift was also responsible for the maintenance of almost all the existing lift units in the GDR. Even more than the relatively good condition of the enterprise, it was access to the maintenance contracts that made VEBLift interesting for Western investors after the fall of the Berlin Wall. By the end of 1989, competing Western lift firms were entering into negotiations over a joint venture with the management of the enterprise and with the ministries, which by 1990 were becoming less and less influential. The contract was finally awarded to a large multinational lift firm, which in September 1990 acquired most of the shares in VEBLift. The multinational company Hochinauf signed the contract with the aim of opening up the market in Eastern Europe and securing the maintenance contracts.

When I began my fieldwork in July 1991, Hochinauf owned a production unit in East Berlin and an additional factory and an administrative centre in West Berlin. This is where I carried out comparative fieldwork in December 1992.

Notes

1. *Kombinat*, translated by 'combine' or 'trust' is 'a modern type of production unit in the various sectors of the GDR economy consisting of several juridically independent enterprises and managed by a director general responsible for concentrated and efficient use of all the resources at the disposal of the respective enterprise' (*Kleines ökonomisches Wörterbuch*, Deutsch/Englisch, Berlin: Verlag die Wirtschaft, 1975).
2. Even the Wandlitz villas of the high-ranking SED (*Sozialistische Einheitspartei Deutschlands* – Socialist Unity Party of Germany) party officials, which created a lot of talk shortly after the fall of the Wall, were distinguished more by petit bourgeois affluence than by real wealth. In contrast to the Soviet Union, where in the era of the planned economy considerable fortunes could be accumulated, the GDR seems economically to have been a far more egalitarian state.
3. Translator's note: *Wessis* is an informal term for West Germans and *Ossis* for East Germans.
4. In the reforms of the New Economic System (1963–70), cybernetics was used to strengthen central planning (Meuschel 1992: 188), as the subsystems were supposed to be self-regulating. The enterprise was seen as the catalyst for technocratic-systemic self-regulation that was to be integrated both politically and materially.
5. In a capitalist company, this would correspond to a department building labour-saving devices.

Part I

Memories of the Planned Economy

In May 1990 the employees spoke about the planned economy in a way that reflected their preconceptions of the market economy; that is they spoke about the conditions during the period of the planned economy in a way other than they would have done before the fall of the Berlin Wall. In this way, employees at all levels of the hierarchy emphasized the aspect of informal economic activities outside of the plan, while the dominant GDR discourse about the scientific basis of economic planning, disappeared almost completely from the accounts.

Employees conveyed an image of themselves tailored to the role that they wanted to have in the new social system. The accounts of the past were statements about the present they were experiencing. Their images of the past served either to legitimize the present social order or to question it. Conversely, the perception of the present was largely connected to the experiences of the past (Connerton 1989: 2). During my periods of research in the enterprises between 1990 and 1993, the people I talked to were constantly obliged to adapt to new conditions in their private and professional spheres. The preconceptions they had formed about the future of their enterprise and the chances they saw for their own professional career were continually superseded by the economic and social developments and had to be rectified. The further the German unification progressed, the more the personal relationships in the 'people-owned enterprise' (volkseigener Bertrieb, VEB) were depicted as a mirror image of new experiences in the market economy. The discourse about the past became part of a systematic comparison between East and West and from the end of 1990 it was this comparison more than anything else that caused a stir.

At the same time, the planned economy and the power relations that dominated it belonged increasingly to the past, while reports about a lack of discipline, semi-legal transactions and a lack of work motivation became more and more outspoken. Some aspects of budgeting in the planned economy, which before were only discussed in hushed voices inside an intimate circle or were rumours, were now related to me extensively and in detail. Just as revealing as the relations of the past, which were now reported openly, were those that were kept silent or no longer spoken about. In this way, it was not only the accounts given of the past that were relevant, but also what was not revealed, because it was considered trivial, or important but too intimate. Events and circumstances, which the subjects had felt to be hurtful and shameful, were sometimes forgotten or omitted in the accounts.

The political view of the interviewees towards the planned economy and their chances in the market economy influenced the accounts they delivered of the past. Devout party members guarded their memories like treasures and were not prepared to grant me a glimpse into the brigade diaries they had kept. The books confronted their keepers with their past political statements and were, depending upon their present conviction, either destroyed, collected or hidden.

For the reconstruction of social relationships in the enterprise during the time of the planned economy, it was important that the members of groups, who had worked together in the past, could still directly relate to one another. Other members could then supplement, correct or confirm their accounts. Political views and actions of colleagues, who were opponents in the past, were discussed in detail, only as long as they were still in the enterprise and had some influence. Defeated opponents were no longer of interest once they had left the enterprise and few stories circulated about them. In those groups that had remained relatively stable since the period of the planned economy, the memories were more varied, had many more facets, and manifested themselves in all their inconsistencies.

Chapter 1

Ideology and Practice
of the Plan

Progress is what we call something
That often only creeps away.
If we called it sneaking away,
That would be O.K.[1]

Horst Froberg, (alias *Fröhlich*, skilled worker at Stanex, unpublished poems 1988)

The transfer of private ownership of the means of production into 'people's property' (*Volkseigentum*) did away with profits on capital and the maximization of individual gain as the primary drives of the economy. Therefore, the market lost its function of allocating economic resources through competition among profit-oriented, legally autonomous subjects, acting at their own risk – in the roles of sellers and buyers of labour power, investors, producers, dealers and consumers (Rottenburg 1992: 242). To replace the market, the plan was introduced by means of which the appropriate products were supposed to be manufactured and distributed in the required amounts and at the right time.

The architects of socialism assumed that economic and social relations could be planned and that such planning would bring large gains in productivity and efficiency as opposed to the 'chaos of the market' (Kornai 1992: 110). However, Weber had already pointed to the fact, 'that there can be no talk of a rational "planned economy" as long as there is no efficient calculation system for rationally setting up a "plan"(Weber 1972: 55–56). Classical economists like von Mises (cited in Hirschhausen 1994: 16) maintained: 'The planned economy is not an economy at all. It is merely a system of fumbling-around-in-the-dark.' In the planned economy money had lost its function as a general means of control, and therefore a different system of valuing economic activities and comparing them with one and other was needed.

In the planned economy the scope enterprises had for making decisions was curtailed considerably by production planning, political prices, and the central allocation of production materials. On the other hand, the enterprises did not have to bear the financial consequences of their decisions. The people-owned enterprises operated under 'soft budget constraints' (Kornai 1992) and could, without any financial consideration, hoard goods and workers, which another place in the system then consequently lacked. Since goods, spare parts, and tools could not simply be purchased with money, the inadequate supply in the enterprise had to be compensated by 'self-initiative' (Adlerhold a.o. 1994: 33). Although the system had portrayed itself as a centrally controlled planned economy, in practice it was largely dependent on informal autonomous patterns of action of the economic agents. After the Berlin Wall came down, the employees in the enterprises I spoke with emphasized these patterns.

> If you read the plans carefully, it definitely states that the production plans, the yearly plan and derived from that, the planning target of that year, that all this is law. So in principle it is like the law, although nobody – hardly anyone – bothered about what it said. That's what it's like, that's the decisive factor. (Ruland, skilled worker, Stanex 18 April 1991)

All the accounts about life in the enterprises during the time of the planned economy stressed the informal aspects of economic activity. They remembered that employees working in materials management became crafty dealers, who avoided bureaucracy and skilfully obtained the materials necessary for production. Those working in sales explained how they made many things 'possible'. Accountants described how they would skilfully manipulate figures for the plan, adapting them to suit the wishes of the inquisitive institutions. Among the directors, the motto was, 'a socialist director without connections is like a capitalist without capital'. Socialist workers were reknown for their ability for improvisation. The accounts emphasized the innate distance from the official version of the planned economy that came from real-life experience and implied that the narrators were thus equipped for market economy.

However, this description concealed the fact that central planning moulded all aspects of the economy. The implementation of the central plan, this monumental work of bureaucratic coordination, as Kornai (1992: 114) calls it, tied the central planning commission, the planning departments of the ministries, Kombinate and enterprises all closely to the organs that executed the plan. Even the efforts to avoid the plan were inseparably linked to the mechanism of planning. Yet, this raises the question whether informal economic management directed at getting around the plan was contributing towards fulfilling the plan, which almost all enterprise directors unanimously claimed, or whether this in fact sabotaged it.

Planning Scarcity

In the planned economy the producer was king. Attempts in the 1970s to adapt the plan increasingly towards consumer wishes succeeded in taking basic needs into account, but were insufficient when it came to satisfying new or changing demands for goods or services. The planned economy remained right to the end tied to a production plan which was devised according to political targets and which only secondarily was oriented towards consumer wishes. Due to planning from the top down, it was impossible to ascertain the relative importance of demands for various types of goods. The party organs claimed to be better at finding out the needs and fulfilling them than the workers and the consumers themselves. In agreement with the socialist ideals of equality, they had in the first years of the GDR a minimalist definition of the fundamental needs that were to be satisfied. The regime prevented people from consuming by providing few goods and it claimed at the same time that living standards under real existing socialism would constantly improve.

State enterprises in the GDR kept a record of material costs and labour, as well as of general expenses. However, the prices which the consumers were being charged and which were counted when the plan was being balanced, were not simply derived from production costs, nor were they market prices. They were political prices calculated to influence the demand of the population in a way that was politically appropriate (Kornai 1992: 153). Enterprises producing consumer goods counted toward their plan fulfilment an enterprise price plus a production tax – a type of luxury tax that was fixed by a price commission and deducted directly by the state.

In May every people-owned enterprise committed itself to a higher production figure than in the previous year, while requesting a correspondingly higher share of materials. In the ministry, a state planning commission then checked if planned production and needs were in concordance with one another. Production was, however, always less than the declared needs for materials and parts. As the demands for materials would never be approved completely, enterprises asked for more than they did in fact require. Employees in materials management ironically called this process 'Mittag's balance sheet ideology',[2] which was a consequence of 'unavailable production', or, in other words, of permanent shortage.

> If your enterprise needed three parts, then you applied to the Kombinat for nine parts. The Kombinat reduced this to six parts and then applied for six at the Economic Affairs Council. And they reduced it all again to three and then you got your three parts. (Kabel, personnel resources manager, Taghell, 23 April 1991)

In January fine planning ensued on the basis of the state-planning task (*Staatliche Planaufgabe, STAG*), in which the state established the amount to be produced. The actual materials delivered by the state in response to the requests of the enterprises were never sufficient and they did not correspond exactly to the sub-items on the list. The foremen in the cutting department of Hochinauf, for instance, complained that they had received three millimetre steel sheets instead of two millimetre ones, which cracked when being pressed and increased the weight of the lift cabin.

Enterprises received less of some materials and more of others – more in fact than they actually needed for production. They were able to hoard materials that other enterprises needed urgently. To force enterprises into making use of the stocks they had accumulated, they had to present a concept for stock utilization if they wished to apply for new materials. The head of materials management at Stanex explained how he found a way round this clause. If the enterprise, for example, required one hundred thousand capacitors of a certain type – all capacitors circulated with the ELN number[3] 13.272 – and it still had 30 thousand capacitors of the wrong type in stock, it would only be granted a quota of 70 thousand capacitors, which would not be enough for production. Therefore in the planning phase materials management declared a requirement of 130 thousand capacitors and invented bogus products for these so that the surplus would not be noticed by the plan inspection.

The planning authorities tried in vain to prevent and to monitor such abuse of the system by demanding increasingly detailed supplementary balance sheets, such as concepts for stock utilization. In addition to this came a detailed tendering of account statistics as well as reports every month and every ten days that were checked by the Kombinat and the Economic Affairs Council of the district. The sheer amount of collected data in fact made it more difficult to monitor the planning since they lacked the technological means – such as computers – to organize the data and analyze it.

Investments in production machinery had to be requested years in advance. Yet, even if the requested machinery was supplied to the Kombinat, the enterprise that made the application could not be certain that it would receive the requested item because another enterprise of the Kombinat with better connections to the Kombinat directorate might snatch the machinery for itself instead. This was a scenario faced by the Taghell enterprise again and again.

> The Kombinat director was also the director of the main enterprise. And from all the quotas that arrived, he first of all took the biggest slice for himself and then handed out the rest. And this distribution was in turn dependent on how firm a director could be towards the Kombinat director, how well he got on with him personally. From time to time, he had to get some present from our own director … . (Kabel, personnel manager, Taghell, 23 May 1991)

At the time of the Wende, the machinery of all three enterprises was fifteen to twenty years old on average. Taghell had the oldest machines, some of which dated from the pre-war period. When I visited the prefabrication workshop for the first time in June 1990, three of the stamping presses stood in pools of thick oil, and the sixty-three ton Romanian eccentric press pounded out of control. Both of the latest GDR machines stood around unused. The enterprise had acquired them in 1988 at four times the normal price, as they had originally been intended for export to the West, but both had to be recalled due to technical faults. Along with the oversized dimensions of the machines, it was these faults that rendered them useless for the enterprise. For ten years, the production unit resorted to using a thread roller they had put together themselves out of an old lathe, on which the thread for the sockets was rolled. Repeatedly the enterprise made an investment request for a new roller and although this was approved, the desired machine never came. An attempt to buy a new lathe instead and convert it, came to nothing because the converted machine did not reach far enough for the necessary thread depth.

Machinery at Stanex and at VEBLift was in 1989 behind the times, however, VEBLift had been granted the chance to obtain some new GDR machinery as a result of a state support programme for new housing, including blocks of flats with lifts, at the beginning of the 1980s. Stanex on the other hand looked after and repaired its old machinery and could rarely get its hands on a new piece of equipment via the official channels. These examples of incoherent investments clearly show the permanent conflict between the productive, social, and political functions of the enterprises (Hirschhausen 1994: 8). In the people-owned enterprises, the very latest technology stood side by side with machinery that was ready for the scrapyard, elements in the production chain were often missing and the technical design of the machinery was unsuitable for the production requirements.

Alongside the official planned distribution, which was useless for fine-tuning production planning, there were two other lines of distribution: political-bureaucratic distribution that was based upon personal relationships along party lines, and pragmatic redistribution that encompassed the direct exchange of urgently needed investment and consumer goods between the firms and individuals. An official form of this exchange of goods was the 'exchange of equivalent performance' (*äquivalenter Leistungsaustausch*), which permitted departments that manufactured means of rationalizing production to exchange these with each other. Stanex interpreted this arrangement in such a way that it declared components produced by other enterprises as a means of rationalizing production and exchanged them for machines and concepts. Although there was officially a rule of precedence in the waiting list for machines set by the Kombinat, customers who were not at the top of the priority list, or only wanted to produce small quantities of goods, still received a machine if Stanex needed something in exchange from them. In the year prior to the unification, the enterprise set up an exchange with a manufacturer of milling machines, which delivered milling heads, and received project documentation in return. Entire machines were produced and exchanged without leaving any trace in the plans. In this way, Stanex produced production machinery for a firm that did not really need it at all, in order to exchange it for a precision milling machine.

The enterprises reacted to the permanent shortage of materials and spare parts by exchanging with other enterprises and by setting up personal networks. As money lost its value as a generalised means of exchange, barter transactions took place that made both sides personally dependent upon one another. In the grey economy, the 'value' of a product was determined by its importance as an investment good or by its popularity as a consumer item. Together with other enterprises, VEBLift organized an exchange of materials, which was tolerated by the system. They came from material warehouses, which the enterprises had been able to build up as time went on. The goods that were exchanged there were often no longer officially declared and did not appear in any of the statistics.

Employees working in material management told again and again how they loaded up the trailers attached to their Trabant cars with bananas, peppers and other items that were difficult to get outside of Berlin. Supplied with these bartering tools, they would then scour the producers or resource managers of other enterprises for materials or spare parts they were short of. The manager of the pre-production department at Stanex, Grabher, vividly described how he personally went off to the delivery firms if they were short of a special part for the production of the roundtable assemblers. If he was told they did not have the part in stock, he would take the woman from the administration of inventory to one side and let her have a look in his trailer. He would sell her things that she wanted for her own personal use and in return she would find him the part Stanex needed urgently for production. In this way he developed a network of relationships over the years that he could fall back on without trouble. Along with obtaining materials for the enterprise, he tied in a private trade exchanging goods that turned out to be a not altogether unprofitable venture for him. His skill was indispensable to the enterprise and won him respect from his colleagues.

Belonging to social networks was crucial to compensate for the restrictions in the supply line. As there was often only one producer of a special item in the GDR, relationships to this producer were critical and had to be maintained. In modern Western industrial societies money is used to objectify the relationships of exchange and to abstract direct personal structures of dependency. The planned economy created new personal structures of dependency, which limited individual liberties (Adlerhold et al. 1994: 34) but which, from within the system of the planned economy, were perceived as liberties and forms of refusing to conform.

The difficulties recounted by the interviewees show that producing scarcity was already pre-programmed when the plan was being drawn up. This was aggravated further by operative difficulties. The warehouses often had long turnover times and were in a chaotic state. A Stanex employee coined an expression for it, 'the most common products in the warehouse were the NIWs (Not In Warehouse)'.[4] Materials disappeared during production because elements for electronic construction, for example, were of great interest to do-it-yourself men, who were putting their own electronic devices together at home.

Fully meeting a production quota set by the plan could not be guaranteed despite bouts of intensive labour. In order for the enterprise to claim completion of the plan, quotas had to be corrected and/or manipulated. The balancing of the plan offered a few possibilities for this. The type of goods that were to be produced was specified. However, producing the right type of good was less important than complying with the reference number for the industrial production of goods.

> This reference number does not exist anywhere else in the world, this reference number for the production of goods. Behind it is always this concept of fulfilling the plan. And this production of goods was also invented by socialism when the economy of socialist enterprises was invented. (Grabher, production management, Stanex, 7 May 1991)

The reference number of the industrial production of goods corresponded to the value that every enterprise could credit itself with in the balancing of the plan. This number was neither identical to the retail prices, nor to production costs, but depended upon the amounts produced and on the export bonuses and luxury taxes. This allowed the enterprise to juggle quantities and the amounts credited towards the plan. Products that were judged by the product commission to be luxury items, like the brass lamps Taghell manufactured, were taxed with a luxury tax, the so-called production tax (*Produktionsabgabe*). This production tax was added to the production figures for the final balancing of the plan. It was therefore advantageous for an enterprise to manufacture a relatively small number of very expensive items, for which a high luxury tax had to be paid, and supplement them with a high quantity of cheap goods that were easy to manufacture. Taghell produced very elaborate and expensive brass lamps in order to be certain of completing the plan quota in terms of value. Taghell then supplemented these lamps with a high number of inexpensive tin lamps that were easy to produce but for which there was almost no demand.

The wishes and needs of East German consumers were secondary in these calculations. The brass lamps, which the state taxed as luxury items at a particularly high rate, were supposed to exhaust the purchasing power of the GDR citizens and satisfy their desire for luxury goods. The demand for the heavy brass lamps in the style of bourgeois living rooms at the turn of the twentieth century was enormous. Despite the high prices (1,200 marks for a five-armed hanging chandelier demand could not be satisfied. The sales department at Taghell was always busy fending off customers. Although prices for consumer goods had to be confirmed by a central price commission, it was still possible for the enterprises to manipulate these prices. Since the enterprises were obliged to present innovations in their product range every year, their prices for a 'product of improved quality' was confirmed without a great deal of extra trouble.

If the enterprise, mediated through the state department for foreign trade, produced goods for non-socialist countries, the credit was four to five times the amount it would have been if the item was sold domestically. Taghell fulfilled special contracts for government institutions and party members in addition to the usual orders for the domestic consumers and for customers from non-socialist countries. The party and the government operating outside of the plan ordered lamps for prestige buildings such as luxury hotels, ministries and guest-houses belonging to the party. The employees highly valued these special orders for in these cases they received higher pay and were provided with all the necessary materials without any problems. The department manager who was in charge of these special orders explained to me that he was always a welcome guest in the hotels that he had furnished with lamps.

If producing investment goods, and above all so-called means of rationalization, the enterprise could practically set the prices at will. The design engineer Kater relates:

> In the GDR there were fixed prices for all products, but not for special machinery. We made our prices ourselves. There was nothing to it of course because as everyone wanted our machines, we set the prices so that we always exceeded the plan. (Kater, design engineer, Stanex, 20 February 1993)

Although officially the prices of production goods had to be confirmed by the price commission, it usually accepted the calculations of the producers (see also Kornai 1992: 149). The Stanex department of assembly automation could, for example, double the proportion of overhead costs in the price of the product in order to increase the profit margin, with which it could then balance out deficits in other departments of the enterprise. The further development of the roundtable assembler enabled the standardization and rationalization of production. Yet, this did not lead to sinking product prices as it might have done in the market economy. Instead the improvements in the quality of the machinery were brought forward as a reason for raising prices.

The purchasing firms often welcomed the higher prices because they allowed them to fulfil the annual requirement to invest in means of rationalization. The national statisticians could then celebrate the capacity for innovation of the GDR economy, without a corresponding modernization of production machinery. What in the people-owned enterprise was called 'earning money', only had a remote similarity with the mediation in the market economy between production and demand through money. The producing enterprise could not freely use the sums of money it had been credited with, nor did the purchasing enterprise have to create financial reserves for the 'purchase'. Any money earned was merely written on paper and could not be freely converted into merchandise.

'Money' and 'tax' are two words that in a market economy seem to have intrinsic meaning. However, in an economy of scarcity money did not represent wealth. Accumulation of large amounts of money was not sufficient to improve one's lifestyle. Barter and or personal relationships were necessary to gain access to desired goods or results. Money functioned only at a minimal level to fulfil basic needs as defined by the state. Its role as a reservoir for future consumption was limited. Also, product taxes had different implications in real existing socialism. In market economies a product tax is a levy that finances the operations of government. It is unwelcome by most capitalist enterprises as a political instrument that negatively influences demand. In the socialist economy of the GDR it took on a different and dual nature. It was mainly a pricing mechanism to curb demand but could also be a credit to the enterprise. It became a favourable factor in the calculations used to determine whether an enterprise had fulfilled its obligations under the plan. While producing goods was a strain for the enterprise, luxury taxes were a benefit, and being allowed to produce luxury goods a privilege.

Politics before Economics

Increasing productivity and reducing the workforce had been officially declared in the 1970s the highest priorities for the development of the GDR economy. Erich Honecker demanded in 1979:

> Now it's a question of raising the tempo of intensification and getting results on the scale of the national economy more quickly. … More national income, by means of a better material economy, of more thorough use of working time and production facilities – our economic policy is aiming at this. (Honecker 1984: 265–6)

In the 'competition of innovators', the employees received financial bonuses and symbolic recognition for new ideas for improving work efficiency. The departments constructing means of rationalization of the Kombinat and the enterprises were encouraged. These departments had to manufacture more and more appliances, machinery and installations for their enterprises to replace worn-out means of production and extend production facilities. The value of the in-house production of investment goods rose in the GDR from 0.8 million marks in 1975 to 9.3 million marks in 1987 (Grünert 1997: 103). These measures were to improve the technological standard and raise the level of productivity, without however touching the existing structures of employment and production. In the eighties, the production and employment structure developed in a direction that was diametrically opposed to developments in Western industrialized nations: increasing instead of reducing the vertical integration of manufacture; widening instead of reducing the product lines, diminishing instead of increasing the division of labour between different enterprises (Grünert 1997: 106).

Larger individual initiatives were unwelcome in the GDR economy. The bureaucratic structure imposed itself by bundling incentives for changes in production and for new products through the political organs and then passing them on to the production basis via strictly predetermined channels. Rudolf Bahro succinctly summarises this mechanism:

> The mechanism of 'task distribution' to individuals is moulded by the bureaucratic-centralistic form of planning, in which those at the top preferably only receive 'questions' and passive information about how the current situation is, whereas they deliver active information about how the situation should be. According to this principle, people do not themselves look for tasks, or recognize and tackle problems, but are obliged to have these appointed to them. (Bahro 1977: 252)

Stubborn adherence to their own strategies of problem solving, even if they were conceived for the 'good of the people', could also be understood as implicit criticism of the official line.

By the end of the 1970s, the systematic prevention of any individual initiative led to economic and social stagnation. Although the employees did not yet understand this stagnation analytically, they certainly sensed it.

> In the seventies I had the feeling there was still positive economic development then, even with respect to consumption! … Flats were built, nurseries were built, people had jobs, they did their flats up. In the seventies everyone had all they basically needed and then nothing new came along! No new ideas or anything! (Schuster, skilled worker, Stanex, 22 April 1991)

The managers claimed 'economic success' in the 1980s just as they had before. This success was measured by regular fulfilment of the plan, which was achieved by skilled manipulation of the plan criteria, by production for the non-socialist hard currency countries, by work on a production line guided by the priorities of the state, by development of new products and above all by realization of new investments also outside of the plan. This economic success was the most important justification for the power of the socialist leaders both inside and outside of the enterprise. National economic considerations contradicted this and competed with the economic strategies of the enterprise directors.

> There was the illusion that they should think on the level of the national economy. … The most striking thing was of course that the comrade director never thought on this level, at least those I know, but they always thought in terms of the enterprise. Since the correction factor of competition was missing, the way of thinking in terms of the enterprise economy could therefore catch on fully to the disadvantage of the national economy. (Kater, design engineer, Stanex 25 April 1991)

Attempts at further production in the interests of the national economy, like the section director at Stanex attempted with the development of the round table assembler, met with little approval among the superiors in the people-owned enterprise and in the Kombinat. They saw it as creating new problems. Materials had to be obtained and plan quotas for their enterprise might increase. This stance was widespread throughout the GDR. As Harry Maier writes in his analysis of the inertia towards innovation in the GDR planned economy, 'the socialist production units' shied away from

> radical innovations like the devil avoids holy water. One reason for this is because radical innovations were very risky and painful sanctions could be expected if they went wrong; another reason is because, if the innovations were at all successful, rapid increases in efficiency would be possible and could only be repeated with difficulty in the following year. (Maier 1993)

After the Wall fell, it was with relief that the workers and employees in the people-owned enterprises welcomed the disappearance of the numerous accounts and quotas for a plan that never kept to what it promised. Immediately following its collapse in 1990, party supporters and opponents described the planned economy as a magnificent pipe dream that had been built again and again at massive expense of time and labour.

Yet the criteria of plan fulfilment remained a point of reference in the interpretation of the real socialist past and in the assessment of individual importance for the enterprise. A positive discourse about personal engagements for fulfilling the plan was present among most employees of materials management and many members of the production brigades. Members of successful brigades never missed the chance to point out that they managed to fulfil the plan in spite of all adversity.

Since plan fulfilment challenged people to contravene the rules of the planned economy, it meant that even those who managed to fulfil the plan could potentially become guilty of transgressing it. This lead to decision makers seeing themselves obliged to take action and argue constantly in a sort of double logic. On the one hand, they felt officially obliged to defend the principle of the plan as if it were law, yet, on the other hand, they knew about the shortcomings of the plan and attempted to fulfil the plan using means that were beyond the plan itself. Making the unachievable into an obligation made the ideology of real socialist planning take on an irrational, almost religious, character.

Analogous to Christian belief, where one must essentially believe in primal contradictions like the resurrection of the dead or the trinity of God in one person, employees in the GDR were expected, contrary to better knowledge and despite their daily experiences, to believe in the rationality and the superiority of the planned economy as opposed to the market economy. To express disbelief or doubt openly was a sin and had to be punished. Bahro (1977: 288) describes the structure of GDR society as 'quasi-theocratic'.

> With the presumption of knowing the laws of history and the true interests of the masses, every decision can be justified despite its huge cost to the economy. The 'primacy of politics' where there is a monopoly over the forming of political opinion, involves the fact that factual arguments do not count especially for the highest expenses. (Bahro 1977: 288)

According to official ideology, the planned economy of the GDR defined itself in competition with the market economy of West Germany, which in the 1960s they wanted to surpass, but in the 1980s merely wanted to catch up with. The official ideology was riddled with Cold War logic. The economic agents should be 'mobilized' in the struggle to obtain an economic equilibrium between the military blocks. Still in 1988 the annual motto for socialist competition ran: 'my place of work – my battlefield for peace'. The aim set by the prefabrication collective at Taghell in the socialist competition of 1988 expresses this position:

Referring to the decisions on nuclear disarmament between the Soviet Union and the USA, it is the duty of every-one of us to contribute even more to the creation of an economic equilibrium between the socialist camp and the imperialistic powers through setting ourselves even higher targets. In the longterm, the military and political equilibrium can only be maintained by simultaneously creating an economic equilibrium and therefore our job is our battlefield for peace. (Taghell, author, foreman Saller, 24/3/1988)

This appraisal of the planned economy agrees with the analysis by Sapir (1992), who sees the planned economy as a permanent war economy emulating institutions that managed shortages during the war. This economy, fixing quotas and prices, was suitable for mobilizing and rationing resources to enable the country to survive extreme situations. As Honecker himself said, this was 'in order to overcome the chaos left behind by Hitler's regime, in order to build up the economy for the benefit of the people and to make it competitive' (Honecker 1984: 68). Keeping the idea of a strong enemy at the forefront also allowed the regime to explain away the difficulties of the planned economy. Continued shortages were then acts of sabotage by hostile forces. Driven to the extreme, that also meant the employees were called to do their best in order to defend their fatherland and socialism. Reluctance to work became treachery against the fatherland and took on a political dimension.

At the same time, the ideology of the mobilized economy also provided justification for economic management that circumvented the plan. Fulfilment of the plan – equated with defending the fatherland – was the highest maxim and the means of achieving this aim then came second. In this way, the production management of assembly automation at Stanex could proudly call itself the 'Blue Light Collective'[5] because the employees succeeded repeatedly in obtaining spare parts for the production machinery or components for machine production. Informal economic management in the interest of the enterprise and of the pragmatic preservation of the functions of socialist economy was highly valued. Enterprise managers exchanged with one another materials that were necessary for production and placed orders with each other that never appeared in the balancing of the plan. Members of political organs received presents and favours, in order to influence decisions about investments on the superior political level. Even party secretaries were proud if they could acquire materials for their enterprise via party connections. Whenever political organs issued special commissions, the enterprises involved were supplied, outside of the plan, with all the necessary materials, tools, and were allowed to pay higher hourly wages.

These forms of parallel economic management were recognized and tolerated – because the continued existence of the planned economy depended on it – but they were hard to check because of their informality. It frequently bordered on illegality. If, for example, the exchange of hoarded production materials no longer went through the books, there was a big temptation for those involved to pocket some of the proceeds. The employees at Taghell told stories about carloads full of lamps vanishing into the night, about members of the party and the government who helped themselves in the lamp warehouse, and about company directors who officially scrapped valuable stocks of brass in order to sell them unofficially.

The amount of truth in such stories is hard to verify, just as it is difficult to measure the extent of illegal transactions and personal profiteering that went on. Yet it was the observations and gossip among the workforce that kept these practices within limits, which also, however, had the effect of creating secretive insider groups of those 'in the know' both inside and outside of the enterprise.

Although manipulating the figures of the plan came in useful for some enterprises for fulfilling the plan, it also increased supply bottlenecks in other places in the system. Whereas the produced amounts and values approached the planned figures, larger and larger gaps opened up in product quality. The quality of the products did not conform to consumer needs – the simple tin lamps from Taghell did not find any buyers. Although the informal acquisition of capital goods, spare parts and materials assisted in easing the heavy-handedness of central distribution, it made the system increasingly difficult to fathom. The people-owned enterprises that managed to get hold of high-quality technological investment goods were often not those who really needed them or could make full use of them. In contrast to what some of its protagonists claimed following unification, informal economic exchange between people-owned enterprises – at least in the GDR – did not pursue any criteria of the market economy because there was no free choice between different suppliers and buyers. There was often no access to highly desired goods, unless through some personal connection.

Whereas the plan as an instrument of controlling the economy was manipulated and evaded, it remained an idea that regulated society and controlled the individual. Fulfilling the plan was the official social maxim and norm, by which the social value could be measured of each economic unit and the people working in that unit. The planning itself was subject to political ideology. State priorities in the planning were founded upon ideological grounds. On top of this, in all the reports submitted before party committees, the ideological work had to occupy prime position, superior to the practical problems involved in fulfilling the tasks set by the plan. The practical economic problems of the planned economy were covered up by political objectives. The plan therefore did not only replace the market and take over distribution of resources, but it was also a political instrument of control. The aims did not correspond with those of economic rationality and the implementation did not conform to the rationality of the political aims. Through political control and the setting of priorities, however, an illusion of economic stability emerged. For some employees, who felt uncertain after 1989 because of the incalculability of the market economy, this illusion seemed in retrospect like a guarantee of future security and transparency.

Notes

1.   *Etwas, was wir Fortschritt nennen,*
     Schleicht oft nur dahin.
     Würden wir es Fortschlich nennen,
     Gäb es einen Sinn.
2.   after Günter Mittag, the GDR minister for the economy.
3.   ELN-*Number: Erzeugnis- und Leistungsnomenklaturnummer* (product classification number).
4.   NIWs (Not In Warehouse): *die NALen* (*Nicht am Lager*).
5.   The 'Blue Light' collective takes its name from the flashing blue lights on police and ambulance cars. Responsible for the logistics of production, they played the role of troubleshooters.

Chapter 2

Pact of Plan Fulfilment

From 1 to 5, or the natural order

Number 1, head of department, chews out
number 2 the foreman: this one then goes
to number 3 the worker and has a go at him.
Then for 4 to 5 weeks, the worker goes off
sick.

Horst Froberg, (alias *Fröhlich*, skilled worker Stanex, 1975)

The ideal 'socialist working person' is characterized by 'social responsibility and the highest sense of duty towards his party, his class and our people', declared Honecker in his speech given on the occasion of the twenty-fifth anniversary of the activist movement in 1973 (Honecker 1984: 69). Socialist morals demanded from the class-conscious worker the will 'to be an active instrument', through which the transcendental will of the party and the objectives of socialism were realized (Gorz 1990: 64). The 'socialist worker' was supposed to strive for fulfilment of social responsibility and not for individual self-realization and personal advantages. By overcoming alienation from work and its products, advocates of real existing socialist ideology claimed to have achieved consistency between the aims of society, enterprises and individuals (Rottenburg 1992: 242). Alienation had been done away with when the interests of the individual coincided with those of the working class which represented the interests of the state.

In the planned economy – as the workers at Stanex, Taghell and VEBlift emphasized – they had a high degree of self-initiative, carried out very diverse tasks and retained the certainty of being of some importance to the enterprise. Had the socialist policy therefore been successful in including the gainfully employed as 'complete persons' in the people-owned enterprises and in reeducating them in the sense of a new socialist consciousness (Rottenburg 1991: 309)? The workers and employees expressly contradicted this and stressed that they were able to develop their skills and initiatives in spite of and against the official enterprise policy in the free spaces and 'pleasant niches' of the enterprise. They saw this free space as a consequence of shortcomings and of the practical impossibility of implementing central planning.

Democratic Centralism in the Enterprise

The enterprise in the planned economy, with the managing director at the top, mirrored the political system of the GDR. The management of a people-owned enterprise was organized centrally around departments fulfilling distinct functions. At Stanex, Taghell and VEBLift, there was a central sales department, a central procurement department, a central auditing department (main bookkeeping), a central management composed of elite party members, a director of economics, who was in charge of the planning, a central department for quality assurance and a health and safety inspector. At an enterprise like Stanex, with a workforce of two thousand, these departments were responsible for areas as different as roundtable assemblers, controls for monitoring liquid levels, electronic appliances and components for the National People's Army, and crystals for computers and specialized machinery for electrothermics. At Taghell, fifteen small plants with 340 employees scattered all over Berlin manufactured household lamps and lamp shades, large pressed tin parts and variable capacitors. With a workforce of 1,100 in 1988, VEBLift produced mainly lifts, but also trough chain conveyors for open-pit lignite mines, and the prototype of an escalator. In order to contribute towards consumer satisfaction among the GDR population, it also made Trabant trailers and garden hose trolleys. This wide variety of products increased the complexity of central administration, making the sections that depended on it extremely sluggish and unwieldy.

The problems of the GDR state bureaucracy and political leadership surfaced again and again in the enterprises of the planned economy. The central administrative departments in the enterprise cut themselves off from one another and attempted to shove the responsibility for their mistakes and delays onto other departments, onto instructions that had been passed on from above or onto guidelines not being followed by those lower down the chain of command. In the enterprise, the directors drew their authority from their role as executors of the production plan. Their political free space and their position towards state institutions depended on the extent of their success in countering the sluggishness of the system, whose ideological representatives they were. An economic director who kept strictly to the ideological principles was protected from dismissal and demotion even if he should meet with economic failure. He was at most transferred from one enterprise to the next. The socialist directors operated with a double logic: on the one hand, they defended the plan along with its political and national economic rationality as if it were law; on the other hand, in their actions they evaded this law so that they could fulfil the plan. This duality and inconsistency continued in the power relations of the enterprise. The claims from the directors to absolute leadership stood in contrast to the negotiated liberties in economic practice. Managers, who had to propagate the official philosophy, relied in practice on the cooperation of the subordinates for fulfilling their tasks (Rottenburg 1992: 244).

The ideology of democratic centralism was illustrated by the arrangement of the conference table and chairs in the offices of the people-owned enterprises. Following the model of Lenin's office in the Kremlin, which could still be seen in the Lenin Museum in Moscow up to November 1993, I observed during my initial visits to the enterprises in 1990 that the managers' desks and the adjacent conference tables were arranged in a T-shape. The design and the comfort of the furnishings signalled the position that the manager held in the enterprise hierarchy. Whereas the long conference tables in the boardroom of the directors of Taghell and Stanex were of polished wood and the chairs upholstered, in the foremen's offices there were only miniature versions consisting of plastic-coated tables and simple wooden chairs arranged in the same T-shape.

During reports and team meetings the manager sat behind the desk at the top and the employees lower down the hierarchy on both sides of the conference table. In this arrangement, none of the employees could look the manager directly in the face. In contrast, he was the only one to have a view of all of them together or to look at each one individually. The seating arrangement underlined the manager's absolute authority and created a distance between himself and the employees. The manager embodied theoretically the collective will of those employees present. The decisions he announced in their name were derived from the superior legal status of plan fulfilment and required no further discussion.

During the daily reports with the director of production, much like the one I was permitted to attend at Taghell in summer 1990, a subtle shifting of responsibilities took place. The foremen of the department for prefabrication, assembly and surface treatment sat at the conference table, together with the heads of process planning, cooperation and the administrator of the vehicle fleet. The director sat enthroned behind his desk. At first, the foremen presented problems encountered in obtaining materials and with working conditions that had prevented them from completing their tasks on time. They turned to the director of production and demanded more spare parts. They insisted on replacements for a burnt-out neon tube in the prefabrication department and for specialized pliers for assembly. Then the main part of the report began. In a loud authoritarian voice, the director of production called up long lists of part numbers, which each employee had lying in front of him. Those present had to report where the parts were in the production chain, whether the materials necessary for production were in stock and whether the business partners they worked with had delivered materials on time.

The atmosphere was similar to that of a classroom where the strict teacher looks for those who might be playing some kind of trick. The declarations of efficiency and conscientiousness were commented on ironically and questioned half-jokingly by the colleagues. The director of production tried hard to keep his monitoring serious and sought to create the impression he had his eye on every detail of the production process.

After the report, the foremen explained to me that the figures they played around with during the report never corresponded to the facts. The foreman for prefabrication described how he tried in every report to lower the number of parts his department was supposed to produce and at the same time to produce parts in advance, thus accumulating a stock of finished parts, which he could then fall back on whenever there were bottlenecks in the supply.

State and party officials, enterprise directors and employees played off their interests in the enterprise against each other. The officials, represented by the party secretary and full-time trade union official, had the task of carrying out the social and national economic priorities in the enterprise. The main priorities were the completion of the production plan and the expansion of political and national economic campaigns like the movement of innovators and activists. These campaigns, which were addressing the workers directly to encourage them to achieve a higher productivity, were not particularly liked by the directors. Stakhanovite thrusts in the workshops and on the building sites, though these hardly continued to occur in the 1980s, caused the largest problems for enterprise management. They used up the material reserves and put a great deal of strain on the old machinery in need of repair.

The state and party officials tried via 'social movements' to create alliances with the workers, and in some cases even against the enterprise directors. The official ideology emphasized the unity of interests between the workers and the party. But in practice workers and enterprise directors rejected increased plan quotas and together made a stand against the leadership of the state and the party. Limiting the amount of work undertaken was in the interest of the whole enterprise. A clear rise in production would have led to a rise in the initial planning targets for the following year and entailed difficulties for fulfilling the plan in the future. In tacit agreement with the directors, the workers could therefore cut their work performance down to the level sufficient for completing the plan. To cut it down any further, however, would have endangered the social and political position of the directors. Although the directors operated under soft budget constraints (Kornai 1992) and could renegotiate obligations set by the plan, as well as investment needs and workforce quotas on a political level, they were at the same time politically vulnerable and subject to the arbitrariness of party control if their enterprise failed to complete the plan.

Autonomy and Drudgery

One prejudice that was often heard after unification implied that the natural idleness of workers came to light in socialist enterprises. The workers and employees opposed this, emphasizing how they developed diligence and self-initiative and accomplished this without needing any disciplinary constraint. The women in the assembly department at Taghell insisted that they checked their brass lamps with great care. The men in the prefabrication of lamp components never tired of telling me how they determined their own work rhythm. They stressed their ability to produce independently of the foreman and rely on their own cunning and ability because they knew the production process and kept an eye on it. Workers at Hochinauf underlined the fact that during the time of the planned economy they had to pay far more attention to quality than they did later on in the market economy. How can this self-portrayal of the employees be brought in line with the reality in the people-owned enterprises?

The autonomy of the employees at work was not part of the programme of the real socialist system, but came about unplanned as a consequence of the economy of scarcity. It was, however, also encouraged by the high political and ideological status that was attributed to workers in the 'Workers' and Peasant State', GDR: once employed, a worker could scarcely lose his/her job. Even those who were frequently absent from work could hardly be dismissed, since there was the explicit sociopolitical intent to educate and reintegrate them through work. In cases of serious misconduct, like sabotaging machinery, being absent from work for weeks on end, or stealing large amounts of enterprise property, the enterprise trade union leaders could reject any decision for dismissal or the Office of Labour could put the worker back into the enterprise under the 'positive influence of the socialist brigade'. The enterprises received anyone assigned to them, whether mentally and physically disabled, released from custody or alcoholic, and they could not be dismissed, even if they disrupted the running of the enterprise or damaged the work morale. Colleagues treated them with a mixture of condescension and friendliness, almost as if they were the department mascots.

Jobs were secure, because in the planned economy it was raw materials and investment goods that were in short supply, but also was manpower. A GDR enterprise could be successful if it managed to keep its plan obligations as reduced as possible and to tie up as many workers as it could wrest from the state. Although these workers were not needed all the time, they were necessary to balance out irregularities due to economic shortfalls. The enterprises hoarded workers in order to be equipped for 'urgent tasks in peak times', but also in order to increase the importance of the enterprise in the national economy and the status of its management (Grünert 1997: 55). The enterprises had therefore simultaneously and (from the point of view of the participants in the study) inevitably, both a shortage of labour and endless potentialities for laying off staff (Grünert 1997: 99).

Due to the chronic lack of materials and the insufficient supply of spare parts the rhythm of production continued to be irregular and the production outcome was never completely predictable. At the end of the month or of the quarter in all of the enterprises, 'storming' set in for completing the plan. It required the employees to do extra shifts, put the machinery under too much stress and resulted in poor product quality. Friedemann, fitter in the prefabrication section at Taghell, described to me how he saw the 'storm' at the end of the month:

There had in any case always been a lot of turmoil at the end of the month. Not only with us here, but up in assembly too. At the beginning of the month, we always let things fall behind a little and then at the end of the month there was always a riot. And if you down here had fallen behind, ... then a few people were fetched from the office too ... Then it really got going. Lamps put together, packaged and loads of stuff to do. The end of the month was always a big thing. (Friedemann, fitter Taghell, 29 July 1991)

It was with a certain satisfaction that Friedemann described the chaos of the storm for completing the plan. Divisions between departments were temporarily removed. The office employees had to help out in production and get their hands dirty. He seemed to take the immense stress for granted as it justified the relaxed working rhythm at the beginning of the month. The employees, weary through extra shifts and particularly long hours at the end of the month, considered the beginning of the month a period when they could determine their work rhythm themselves. This rhythm, influenced externally by lack of materials, was reinforced by internal mechanisms. The ability to 'cope with chaos' (Aderhold et al. 1994: 35) placed a kind of bartering tool into the hands of the workers that they could use in informal negotiations with their foremen.

In periods of less activity – of machinery stoppage and bottlenecks in supply – the workers of prefabrication at Taghell disappeared into the different kitchenettes or, if they wanted to be undisturbed, into a nook by the river that lay hidden behind a gate, for which one of the colleagues had the key. When all departments worked hard at the end of the month, the workers of prefabrication pulled along. Their commitment went beyond the obligation to work overtime. They worked right through weekends and did double shifts, during which there were often serious accidents through overfatigue. Workers proudly told me in private how on one of these nights they had used the 63 tonne Romanian eccentric press, which pounded recklessly out of control, for riveting the candelabra arms. This had been very dangerous, but the machine had spat one part out after the other.

The workers found it unpleasant when there were long phases of inactivity at work due to shortages in materials and spare parts. A lot of drinking went on during these periods of waiting. In contrast, in periods of intensive activity alcohol consumption clearly declined. Working at weekends and doing special shifts took their toll on family life, above all when the spouse was also working and the children had to be left in weekend care homes. In taking sick leave, many employees found a way to compensate for all this free time they sacrificed and the long working time of eight hours forty-five minutes daily. From the 1970s onwards, workers' interest in a higher income sank constantly since they could hardly purchase anything with the money they had saved. Even though back in the 1960s it had paid off when employees dragged out the work in the first two thirds of the month in order to earn more in the final third of the month through overtime and weekend shifts, in the 1980s interest was increasingly oriented towards more free time and greater free space within the enterprise (Bust-Bartels 1980). The management had to concede more and more freedom to the employees in order to have them complete the tasks set by the plan.

Since there was a large demand for a flexible workforce that could be employed in a number of different ways, the assembly workers from the final assembly at Stanex were put into prefabrication too. The women at Taghell, who worked threading the cables in, also assembled lamps in times of production bottlenecks. The engineers of the construction department at Stanex went to do special jobs at an electronics production plant at Strausberg, which belonged to a completely different part of the enterprise. The way work was organized resembled what E.P. Thompson described as task-oriented production in preindustrial societies. Periods of doing nothing varied with phases of intense activity (Thompson 1967: 73). Phases of night and weekend shifts followed long periods of machinery stoppage resulting from shortages of materials or spare parts.

The shortcomings of centrally planned distribution and sudden changes in the centrally controlled production policy forced enterprises and their workforces to adapt. Although Taghell had initially produced only simple wooden and tin lamps, in 1978 the enterprise was allocated a quantity of brass and was given the task of producing expensive sets of brass lamps – each lamp costing 1,200 marks – which would satisfy the consumer lust of the population and would absorb their purchasing power. Nobody in the enterprise had ever worked with brass before. As a result, the initial lamp models – designed by employees themselves – were crude and heavy, for they consisted of pure brass. The metal tools available at the time were incapable of cutting the soft brass. The cutting process caused weals and scrapes. Only by trial and error did the workers adapt to dealing with this metal. It was only when years of improvisation had passed and support from a professional designer came to the rescue, that models emerged which were eventually well-received and distributed in their thousands to the West via a Swedish firm.

In the 1970s Stanex developed assembly automates following Western models. The first model had a power motor that vibrated like 'a tractor motor' and hence needed a heavy cast-iron base for stabilizing the structure. Only through gradual further development did the machine become lighter and production series could be standardized. For this development, it was necessary for the design engineers and skilled workers to work in close collaboration, where the less-qualified foremen would hardly interfere at all. For solving some of the problems of detail, the skilled workers had all the freedom.

The three skilled workers who thought up the special vibrating tables, which had to be redeveloped for each machine, were the most independent of all. Every vibrating table bore the individual handwriting of the worker who built it. The three employees, who liked fiddling with the most intricate production problems, were so strategically important to the enterprise and their work so impenetrable that they were at complete liberty to determine the ways and rhythm of their work. They were appreciated by their superiors but at the same time also feared because of their independence. Fröhlich, one of the specialists, described to me his relationship with Voigt, head of production.

> That was in fact always a big concern – for Voigt too – that they could not keep a close eye on us. They would have liked of course to have *their* people on this job. But I think that only lads who are a bit extreme can work here. You just have to approach it without any inhibitions. And that probably isn't even something you can learn. (Fröhlich, skilled worker Stanex, 16 June 1991)

One of the principles of central planning implied that no product within the GDR should be produced by two different manufacturers (Voskamp and Wittke 1991: 19). However, since the supply industry did not function, particularly if the suppliers and the producers were assigned to different ministries, the enterprises and the combines tried hard to be self-sufficient. Any components they could not buy, they developed in mini-series in their rationalisation department. Consequently, they 'invented the wheel' again and again. Exchange of technological know-how among enterprises, called 'socialist exchange' was encouraged but was not systematic and remained more or less superficial. The great innovation and creativity potential of the GDR economy was bound up by inventing products that had already been invented. The employees made substantial technological achievements, but at the same time the technology level of the GDR economy fell behind.

In 1990, the three enterprises of my research owned machinery that on average was twenty to thirty years old. The old technology, which was difficult to monitor, contributed in practice to the workers' autonomy. A capacity for improvisation was necessary to keep the machinery going and to counteract the lack of spare parts and adequate materials. Workers at the Stanex prefabrication, who had to produce special components for the roundtable assemblers that could not be obtained through the official supply channels, reached a degree of accuracy that could have been on a par with modern machines. They could only reach this degree of precision because they had known their machinery and its flaws for a long time and could adapt themselves to it.

A skilled worker demonstrated to me how he was able to mill-cut as accurately as 0.03 mm using a milling machine from the 1960s by relying on experience and estimating it by eye. Most of the skilled workers in the department complained a great deal if colleagues 'poorly treated' 'their' machines in their absence. They used the milling machines to shape groove cutters, which they also judged by eye. When necessary, they produced complete spare parts for their machines. It took a long time to come up with this kind of quality and the workers had a lot of freedom in structuring their working day. Although they had strictly calculated formal work quotas, these could be compensated for by generous set-up times. The workers regarded their machines almost like sentient beings with strengths and weaknesses, with whom one could get along only if one knew them well. The work quality depended upon the maintenance of the machinery, which the workers mostly took upon themselves because cooperation with the maintenance departments was very cumbersome. The readiness to treat the machinery with care depended upon how strongly the workers identified themselves with their task and their product. The qualified workers at Stanex demonstrated more than any of the others how much care they took of their machinery. In contrast, the semi-skilled workers in prefabrication at Taghell completely neglected their machinery. They often did not know the proper approach to the machinery and did not pay heed to safety instructions either. Many of the stamping presses continued to pound out of control because the 'stop' switch did not work. The compressor and rivet machine lost litres of oil every day, which were then simply covered up with sand. A long metal saw, originally designed only for the production of single pieces, was in fact used for sawing multiple pieces, with the result that its casing became fully worn out. During sawing, the metal rods danced around in the guide tubes and made a racket.

Working with old technology was strenuous and noisy, often dangerous and damaging to health. The surface department at Taghell had a real devil's workshop that was only accessible with a key. In this locked room, the polished brass parts were dipped into a varnishing pool at a room temperature of twenty-five degrees and at a constant level of air humidity. The varnish was thinned with nitro-glycerine so that it dried quickly in the air without forming hardened drops of varnish. Although the room was equipped with a ventilation system, this was not switched on in order to prevent the varnish from forming waves on the brass parts. Although other varnishing methods were possible that were less damaging to health, these were not implemented because they would have required the room to be completely free of dust. An installation for burning in the varnish could not be obtained. In the same room, pieces of wood were also treated, which raised the air humidity and exposed those working in there to more fumes. According to the foreman, this work was only for 'certain people'. Both of the workers were heavy smokers and worked for a period of six and a half hours in this room.

Control over the machinery was situated entirely in the workshop. There was no CNC-controlled machinery with remote programming even in the high-tech enterprises like Stanex. Although the management of production in all three enterprises tried to control and monitor the course of production using Taylorist methods, the production developed its own dynamic through the constant necessity to improvise. The workers were not peripheral to their machines; on the contrary, any production using the old machinery in fact depended upon the practical skills and commitment of those workers.

Discipline and Motivation

In the history of work in GDR industry since 1949 there were constant attempts by management as well as by the state and party organs to check and systematize the performances of industrial workers, to reduce free spaces at the workplace, and to put the work under Taylorist discipline. With 'scientific organization of work', an instrument was supposed to have been developed since the beginning of the 1970s that should have theoretically enabled the directors to plan and anticipate the work process (Bust-Bartels 1980: 114). Time norms for certain tasks were drawn up according to either experimental or mathematical methods. In the analytical-experimental method, the norm was established by timing tasks and analysing work during so-called initiative work shifts. The employees had to be informed of the time-taking and were therefore given the chance to perform their work according to the rules. Machinists at Stanex told me that these shifts were specially prepared and that only particular tasks were undertaken with the intention of changing the official norm. During the initiative shifts, they rigidly kept to the task completion times and to safety regulations, which they frequently left out in normal day-to-day business. Sometimes they would also 'inadvertently' break off a cutting bit. The head of production, Voigt, was unable to counteract these tricks. In the end, the changes in norms were always the result of discussions between the foremen and the workers and parts of it were considerably watered down.

Analytical-experimental methods established norms from which time standards for specific tasks were derived mathematically. These time standards were applied each time a similar task was part of a job: for example, if an object had to be clamped in a certain way. These time standards were put together in systems and catalogues and formed the basis for establishing norms mathematically without actually being present at the workplace (Bust-Bartels 1980: 124).

In the Taghell workshop, around 1,500 work routines were attributed with a standard that in most cases had not been changed for years because the technological level had not improved. Sometimes these standards were easy to achieve, sometimes the workers had had difficulties for years. To gain a few extra minutes of free time, the machinery was turned on to the limits of its capacity, while being inadequately maintained. The foreman made unsuccessful efforts to keep the work at a consistent pace to spare the machinery. If the machinery broke down and had to be repaired, this had little ill-effect on the workers' wages, since interruptions due to technical faults or to a shortage of materials meant they still continued to receive a wage corresponding to one hundred percent of the norm. Consequently, stoppages meant more free time in the enterprise.

The attempt to raise productivity among the employees through proposals from innovators often provoked nothing more than improvisation. At Taghell most of the machines, which had been designed to carry out several operations at a time, could never tackle more than one due to technical faults. The workers tinkered with precarious devices in order to accomplish two things at once. For example, on a punch and rivet machine from 1972, the lock devices were covered with window insulating tape, in order to fix two brass lamp arms at once, on which two candle holders were then riveted at the same time. Although this innovation, which incidentally gained the workers a bonus for their 'invention', saved one work step and lowered the norm, it also caused more damage, which then needed a lot of time and effort to be touched up. This was mostly done on the side by the fitters.

The assembly department at Taghell was proud of being named the best department almost every year. The women assembled the lamps according to a norm, which was relatively easy to achieve after some time of getting used to it, whereas the men in the department functioned as foremen, fitters and quality controllers and were not subject to any norm. For every lamp the women finished assembling, they received a brass token that they used for a nightly tally. Prior to the Wende, it was customary to do work in advance and save up these tokens. After a while, some women were able to save over thirty tokens during an average day's production of thirty-two chandeliers. Although officially it was forbidden to leave the enterprise during working hours, the women used this accumulated time to go shopping and to queue for rare consumer goods.

The second type of reward for performance, the bonus time wage, which was above all paid for special skilled work, was toned down over time to a uniform wage. At Stanex, it consisted of a base wage of sixty percent and a performance-based wage for the remaining forty percent. The performance-based wage was assessed according to five criteria: work quality, work quantity, punctuality, adherence to industrial health and safety standards, and dedication to work. The first four of these criteria were in fact fixed. That meant the workers calculated them as components of their wage, which could only be reduced 'as a sanction'. Flexibility in the wage mechanism only kicked in if the worker performed below the normative standard. Only the fifth factor, dedication to work, led to a bonus assessed by the foreman together with the shop steward. For a high dedication to work a bonus of maximum 0.23 marks per hour could be awarded. For an average wage of 6.85 marks, the maximum amount that could possibly be earned was 7.08 marks an hour. In fact the wage system did not reward performance, but only penalised inadequacy.

The weaker and the more undisciplined brigade members could, on a small scale, become a resource for the 'dedicated' productive members of the brigade, as they had to pay sanctions into the fund that was redistributed as bonuses. They were hence tolerated to a certain extent. However, the additional two percent, which could be earned by a particularly hard-working employee through this redistribution, did not offset the loss suffered by the whole brigade if they lost the bonus as a 'collective of socialist work' because of several undisciplined workers. Members of a successful brigade ensured that misdemeanours resulting from poor discipline did not reach the foreman's ears.

The influence exerted by the employees in the planned economy was not based on formal rights of participation, but on their power of refusal. Since management was dependent upon cooperation with the workforce in order to navigate successfully around the rugged rocks of the economy of shortage, the workforce could apply its power of refusal in order to negotiate with their superiors an acceptable work rhythm. They formed a 'pact for fulfilling the plan' (Voskamp and Wittke 1991: 31). This pact was not the result of a peaceful arrangement, but the outcome of practical action, the consequence of refusal and cooperation. The workers took part in the effort towards the end of the month to complete the plan, but they expected in return to be left in peace by their foremen at the start of the following month.

The employees did not have any independent committees who represented their interests officially or voiced their dissatisfaction. In contrast to enterprises in West Germany, where technical-organisational processes of change can be politicized by trade unions and workers' committees (Aderhold et al. 1994: 42), the employees in the people-owned enterprise may well have been able to complain individually, but had no collective power. Open articulation and institutionalized negotiation between opposing parties was impossible. The leadership of the enterprise trade union represented management and party in the work place. They did not possess any organisational or political autonomy. Accordingly, they were not perceived by the employees as representing their collective interests, but as a service organization that regulated the social interests of its members – for example the distribution of places in the holiday camps and villages of the enterprise and trade union.

Even the brigade shop steward, who had been elected by the employees and did not occupy any political function, was alienated in subtle ways from his colleagues. In the case of misdemeanours, he was obliged to countersign the certificate withdrawing the culprit's work bonus and in so doing placed himself on the side of the management. Fröhlich, the satirist and elected shop steward at Stanex, relates:

> As the shop steward, I had to sign this certification. Now if someone really skived off work so much that it was obvious, then I couldn't do anything to avoid putting my name to the certificate so that they received two percent less. Jens often got something or other deducted. Jens, for

example, was working outside on his motorbike or on his car during the lunch-break and then in the middle of everything he left work and everyone knew exactly where he had gone off to. That wasn't very clever, he shouldn't have done that! He could have done that differently. Others also did more or less the same thing, but they were smarter. Well, he made it too obvious. If he gets himself caught then he really has himself to blame. (Fröhlich, Stanex, 16 June 1991)

Fröhlich tried to justify why he had agreed to the pay deduction for his colleague and had helped shoulder the disciplinary measures. His argument was: 'Whoever gets themselves caught, it's their own fault.' The bonus deductions only concerned minor sums of at most forty to fifty marks a month. But his signature, which Fröhlich had to give as confirmation of his colleagues' punishment, took away the autonomy of his role as the elected shop steward of the workers. In so doing, he formally assigned himself to the stance of the management, which otherwise he would criticize.

The organization of work in the planned economy encouraged two typical reactions among the employees, which were not mutually exclusive. The organization stimulated the workers and employees to test the limits of their freedom and to do the minimum work possible. Yet, it also encouraged them to work on their own initiative and to derive their own pleasure from the work and be proud of their product. The workers emphasized repeatedly the joy of a certain 'fool's freedom' in the enterprise.

> Grilling chicken in the tempering furnace, cooking a knuckle of pork and cream whipped on the drilling machinery. But it did nobody any harm. They all continued to work again afterwards. That may seem a little strange to a Westerner, but that's exactly what it was that actually made it all worth living here. We've sat around for ages in this place, we sat in the past from seven till half four. And it was tolerable like that. It was really a great team. But it was the niches that people looked for here that in fact made life pleasant, they were part of the work too. They were not only in our free time. And they were sometimes nice niches, I have to say. (Fröhlich, skilled worker, Stanex, 16 June 1991)

The workers who took part in these tricks and provocations were at other times completely engrossed in their work. For example, when they 'breathed life' into the twelve stations of the roundtable assembler in the final stage of assembly, the two workers responsible, the mechanic and electric design engineers, often worked day and night. The collaboration between skilled workers and engineers, which in daily routine was intensive and independent of formal hierarchies, almost became a symbiosis in these privileged moments. The employees' fascination for technically demanding 'German work of quality' (Lüdtke 1993) hence had not disappeared at the time of the planned economy.

> For us there really is a lot of – and I have to say, luckily for us – individual labour, genuine craftsmanship and also mental work. This was so in the whole section (*Rektorat*).[1] Along with this, the cooperation, starting with construction, technology, work administration and right down to us in the final assembly was all meshed together out of necessity. Over the years this also brought about, directly or indirectly, a community, in which we all helped each other as much as possible. Ninety-five percent of the people did not come here because they wanted a job or had to get work to earn money. So in fact pretty much everyone identified himself with their task and most of them also had fun. And the work would, if it continued to exist, continue to be a pleasure for them. (Ruland, skilled worker, Stanex, 18 April 1991)

The direct non-bureaucratic cooperation between production workers and designers undermined formal hierarchies and the authority of the mid-party elite. The head of department, Voigt, at Stanex therefore tried to bring production politically and professionally under control. As a party member and member of the enterprise militia, he supported the political views of his superior, Dr Schöpf, and showed him absolute loyalty. Up to 1989 he unsuccessfully led a petty war against the skilled workers of his department in order to gain a better grip on the production process. He moved the entire production out of a multi-storey building into an assembly hall, where he could more easily keep an eye on his workers' movements. In order to place limits on the direct cooperation between production departments, Voigt arranged for any work that was being done to touch up mistakes in design or fabrication to be signed with his initials. The skilled workers in the final assembly could no longer go directly to the lathe and mill operators in the prefabrication as before, to get a component corrected or made again from scratch. Voigt indeed managed to reduce drastically the amount of corrective work required. However, he still did not receive effective monitoring of the processes in his department because he lacked the skills to analyse them. The workers in the final assembly delighted in bringing about changes to the machines in conjunction with the designers, but without telling the head of production.

The work intensity in the workshops was based to a large extent on the consensual character of the pact for completing the plan and depended on the interest that the workers had in their tasks. The foremen could assign the workers unpleasant tasks or reduce their annual bonus or in the worst case affect their life outside of the enterprise by writing a negative report if the employees wished to travel to the West for a family visit, but in the end the foremen could hardly force them to work at a higher rate. Since they had few possibilities at their disposal to influence the productivity of the workers, they tried instead to maintain order in the workshop. They tried to ensure that the employees arrived on time for work, did not leave the workplace during their shift and refrained from excessive drinking. The foremen's status in the enterprise seemed to depend on the success of these efforts. They hence emphasized to me repeatedly that they had known how to achieve discipline in the workshop.

The means which Spohr, the foreman at VEBLift, used, were remarkable and recall methods of disciplining from the eighteenth century. He ensured that the workers could no longer observe him in his foreman's office, while he kept them under close surveillance with a Foucaultian panoptic (Foucault 1975: 233ff.). He covered the windowpane of the foreman's office overlooking the workshop with a coat of white paint. He then scratched out three coin-sized holes from the paint at his eye-level – he was about 190 cm (about 6' 2") tall – by means of which he could see out, but the workers could no longer see in. He claimed, however, that these holes were hardly necessary for he could discern the various machines in the workshop from their sounds and could determine if they ran idly, worked in a slow or a fast rhythm. He was thus in a position to keep the workers precisely under surveillance. This form of surveillance recalls an early form of disciplinary society, where discipline had not yet become internalized, 'normal' (Foucault 1986: 241), but repeatedly had to be imposed by those in power. The workers reacted to being constantly under surveillance by refraining from working the moment they escaped observation and they withdrew from the pressure of work by periodically taking long-term sick leave. Spohr's department had the highest level of sick leave in the entire enterprise.

In the prefabrication at Taghell, where primarily a semi-skilled workforce was employed, work motivation was particularly low. The foreman of prefabrication, Saller, developed an almost detective-like flair to discipline the workers and to detect the culprits when discipline was infringed upon. He described the tricks that he turned to in order to create respect for himself.

In order to get some calm in the collective, I drove them into a corner. They would take off from work during the night-shift: they would be there for an hour, then they would be gone! They were at home, that's where they were. That was all very interesting one time in winter … . The fitter had vanished, but allegedly had always been there. There was nothing one could do. One morning, it was more by pure chance, I thought: 'Hang on, you're getting to work at five.' By six they all clock off work. The problem was at that time I didn't start until half six. Yeah, so one morning, into the car and off to work: I saw that all the cars were covered in three millimetres of snow. Only one wasn't. Well, it's logical he had just arrived. All the cars are white. Only his grass-green Trabant wasn't. Well, even the thickest bloke would have seen through that. But he hadn't seen that one coming. He should have shoved his car somewhere round the corner. But they were so confident about the whole thing. (Saller, foreman, Taghell, 21 January 1991)

Nonetheless these disciplinary successes of the foreman were in reality more the exception than the rule. The foreman Saller felt his position in the enterprise to be precarious and thought resignedly: 'Why are the foremen called foremen? Mostly because they are the dummies for men to take advantage of.'[2]

The foremen tried to escape the ideological and material demands of management when possible, but they were not necessarily recognized as 'one of the gang' by the workers and not accepted by them as a consequence. So the foremen mostly sat on the fence, available for those above, but not respected by those below. The foremen Spohr and Saller therefore tried to improve ideologically and politically their position in the enterprise: Spohr, by making his department the champion of socialist competition and Saller, by becoming a member of the party.

The foremen of the 'old school' were the exception there. Foreman Krause from Taghell paid heed in the polishing department to strict discipline and tried to encourage the enthusiasm of his workers for their work through his own good example. His yardstick for what he demanded from his employees was, as he said, the work morale, which he learnt himself previously back in the days of private enterprise when he had done his apprenticeship. In the enterprise Krause had the role of moral authority. As head of the enterprise conflict commission, he openly criticized management and did not refrain from reporting public embezzlement and the trafficking of lamps to the police. His position in the enterprise was untouchable because of the strict discipline and the positive work results he achieved in his department, and his directors feared him for his uncompromising criticism of illegal wheeling and dealings. After the fall of the Wall, Krause was also the first to publicly call the management to account.

The sporadic work commitment and task-oriented work was directed at obtaining a certain result by a certain time without any consideration for the costs and amounts of labour involved. On the one hand, this resulted in workers finding numerous loopholes in the disciplinary system and using these to their advantage. On the other hand, this meant they often remained in the same enterprise for years on end and identified strongly with their work, especially when they became qualified skilled workers. Subsequently, some skilled workers I spoke to had the strong feeling of having had a hand in building up 'their' enterprise and developing 'their' product.

Three factors contributed to the employees being committed again and again to completing the plan. First, the workers and employees gained certain conflict-free spaces through their own personal work commitment, improvisation and flexibility, and this contributed to reducing the rigidity in central planning. In tacit agreement with foremen and heads of department, they cooperated in completing the plan by the end of the month, and expected in turn to be left in peace for the remaining time. Voskamp and Wittke (1991: 32) aptly called this compromise of interests: the plan completion pact formed between the foremen and workers.

Secondly, the prestige of the brigades in the enterprise and the professional pride of the employees depended upon the extent to which they managed to complete their proportion of the plan quota in spite of all the hindrances they had to face. Some employees enjoyed particular prestige because through their improvisation they managed to keep the machines running, to continue production even with faulty or insufficient materials or to succeed in obtaining materials and spare parts. Moreover, 'professional work ethics' (Rottenburg 1991: 317) prevented the skilled workers from feeling unconcerned about their work results.

Thirdly, the employees, who worked hard in the busy periods for the plan completion, were in harmony with the ideals of the reigning social order. Through personal and collective awards, they were confirmed in their ideas of being important members of the socialist society. Many workers in the GDR period were very conscious of their relationship with the state. Thus Mrs Schmidt, an assembly worker at Taghell, felt it obvious that the state owed her something because, 'we too give something to the state. We always worked, and worked hard'. In her eyes, her labour justified the demands she could make towards the state, demands which she knew how to carry out by means of filing official claims and making complaints in the right departments.

In spite of their passive strength (Voskamp and Wittke 1991) founded upon the pact for plan completion, the working class was not the leading class in the GDR. If their passive strength did not cause the efficiency problems of the planned economy and might have contributed to solving some of them, it did not remedy the malfunctioning of the system either. Although the employees could exert a certain pressure by threatening to leave their work, they were neither free to choose which job they would then work in, nor able to rely on independent representative organs to negotiate their working conditions. Looking for work, the position held in the enterprise, and above all the chances of promotion were always affected by personal and political factors, which limited the formal rights of the employees. How this political pressure and control affected the employees on all levels of the enterprise is a topic I will deal with in the next two chapters.

Notes

1. *Rektorat*: Directorate section of the enterprise administered by one director.
2. 'Warum heißt der Meister **Meister**, weil **meist er** der Dumme ist.'

Chapter 3

Socialism as Performance

Thoughts

What is it that torments us like this
When we brood over life
Who is it who shows us the direction
In which we should move?

Why is a person's head so empty
While another's overflows
The one puts his foot down
The other prefers to smile

What is this engine called around here
Where is the destination to fill up
One immediately topples over
The other only totters on

On earth everything is round
That's why one hits an edge everywhere
And whoever does not start brooding here
Does it next door.[1]

Horst Froberg (alias Fröhlich, skilled worker, Stanex, unpublished poems 1984)

Much of the self-confidence the staff was able to muster from the pact
for plan fulfilment was crushed by the so-called 'socialist competition'.
In socialist competition individuals and brigades should compete with
one another for the fulfilment of political goals set for the whole society,
and learn and internalize the classifications, norms, values and roles put
forward by the party (Rottenburg 1992: 243).

Socialist competition was the company philosophy of the planned economy. That is to say, it was not primarily about productivity, but rather about the cultural and political role of the workers inside and outside of the enterprise that granted the system its legitimacy. The worker (in GDR terminology *Werktätige*), as a member of the 'state-supporting class', was a central figure of the official ideology, and the worker's labour was ideologically embellished as 'a contribution towards the wellbeing of the people', and portrayed as 'a contribution in the struggle for peace'. Staff were not only to create goods of material value, but also to reproduce the values of socialism.

Under the political leadership of the trade unions, workers and employees were organized into brigades, which were to participate in the political objectives of the state set by the SED central committee. In a detailed programme at the beginning of the year, the brigades committed themselves 'to work, learn, and live the socialist way'. At the end of the year they accounted for how much of this programme had been successfully fulfilled. Although formally the implementation of the competition was voluntary, the heads of department and foremen were required to accomplish it in an approved form and always with a similar content. The competition year was structured by the commemoration days, which the staff were meant to observe in rallies, brigade diaries and on declaration boards. Finally, winners of the competition could expect a bonus at the end of the year.

After the Wende the competition, which had been regulated differently in the three enterprises, was retrospectively dismissed by most of the staff as nothing more than a farce and an oppressive duty. Analysing it, however, opens up interesting insights into the system of ideological control in the GDR enterprises, as the mechanisms of ideological exertion of power were not ineffective. The ideology subtly infiltrated everyday practices and became a fundamental part of them, as it challenged the staff to adopt a position. In this way, people following the required practices brought about their own domination (Foucault 1987: 254). By being active in the socialist competition, reacting to it or snubbing it, the employees influenced also the actions of their colleagues. The relationship with the official ideology, or rather with what was postulated as the 'social truth', was a dynamic one.

In spite of the ideological power exerted upon them, the employees remained subjects acting with their own free will. They did not lose all chances to resist, even when they expressed their criticism or distance, mostly in secret or encrypted ways (Scott 1990). As Foucault writes: 'Freedom is the condition of existence of power, but it also appears as something that can counteract the exercise of power' (Foucault 1987: 255). If we take this dynamic conception of power as the basis of analysis, socialist society can be uncovered as something other than a decrepit static construction and it is only then that we can begin to understand its mechanisms.

The arranging or resisting mostly took on the form of playing a game, in the sense of putting on a performance. Socialist work morale and loyalty to the party line was portrayed on declaration boards and in brigade diaries, as well as acted out during social events and on the 'stage' of everyday life. The employees-turned-actors had to consider how much of themselves they wanted to put into their representation of state ideology. The forms of representation were varied. They ranged from the perfect imitation of the official discourse in form and content, to subtle irony subverting control. In the 'performance' of socialist competition, the employees shifted between sincerity and cynicism. 'Behind the scenes' (Goffman 1959: 231) the brigade prepared together the presentation of socialist daily life, which was then performed for the official public. Any difference in political opinion and about how to organise the perfect performance remained a secret that the brigade members shared only among themselves. The effect that the fictional world of socialism and its performance had upon the workers profoundly shaped their daily interactions.

Living, Learning and Working the Socialist Way

The emphasis placed on the ideological purity of socialist competition varied from enterprise to enterprise and depended to a large extent upon the political orientation of the directors. The director of the department for assembly automation at Stanex was a devoted SED member and closely supervised the competition. In contrast, at Taghell, internally the competition was hardly monitored but externally it was presented to the party organs in the required form.

In 1990 I systematically asked about the brigade diaries in all the departments of the three enterprises. Yet, most of the foremen and employees who had held the diaries claimed they had thrown them away immediately after the Wall fell, for 'those times had now gone' or 'we don't want to get into any difficulties just because of those things and stand around looking like devoted reds'.

We had to chuck these things away. We said: 'They're going to take us for totally devoted communists if they find our stuff here in the lockers years later. Let's get rid of 'em! Chuck 'em out! No way is anyone gonna tick us off then!' (Born, project planning, Stanex, 4 February 1991)

Other former brigade members, however, doubted whether I had been told the truth in this instance: 'In there are so many memories, and photos and stuff, you can't simply just throw it away. They've definitely still got them. They just don't want to show them to you.' Fortunately, in the economy and sales departments at Taghell I succeeded in photocopying the brigade diaries of the years from 1986 to 1989. The head of process planning at Stanex and non-party member, Grabher, gave me all the brigade diaries which he still owned from 1973 to 1976 and from 1980 to 1985, as well as one from 1987, that is to say, hundreds of pages of material. Grabher commented on them for hours in conversations that I was able to record.

The brigade diaries from the early 1970s were carried out with a great passion for detail. The reports were decorated with drawings and tickets to events. Towards the end of the 1980s the brigade diaries became more factual, were often no longer bound in red folders. Embellishment was limited to photos of brigade excursions and congratulation cards sent by school classes for which the workers had become patrons, or that the women had received for the 8 March, the international day of women's struggle.

> The brigade diary had a defined framework and had to be a standard book. It all started in January, you know, with Luxemburg and Liebknecht who were murdered in 1919. That had to be put in the brigade diary. There was always a big demonstration in East Berlin, right in the middle of winter. Straight after it there were always lots struck down with flu because 150,000 people had been mobilized from the enterprises … 1st of May and 8th of May and the founding of the GDR on the 7th of October and then the October Revolution and so on – the same stuff again and again. These were the official dates that you had to commemorate by doing something special. (Grabher, process planning, Stanex, 7 May 1991)

Despite the required form the brigade diaries had to abide by, the distinct style of each brigade's social life would show through. In 1986, out of the fifteen employees in the 'economy' brigade at Taghell, there were thirteen women and one man who belonged to official organizations: fourteen were members of the Free German Association of Trade Unions (Freien

Deutschen Gewerkschaftsbundes), none of them was a party member and nine were members of the Society for German–Soviet Friendship (*Gesellschaft für Deutsch – Sowjetische Freundschaft*). The authors reported about attending political events and leisure activities, for example, bowling evenings and excursions. They covered whole pages with describing hospital visits and weddings of brigade members. They occasionally mentioned special efforts at work and attached certificates of distinction. The accounts about the activities throughout the year that they had frequently planned two to three months in advance give an impression of their intensive social life as a brigade:

In 1986 the brigade embarked upon the following activities together: on 22 March 1986 they drove to Dresden for the day on the occasion of International Women's Day (8 March). Photos and museum entrance tickets were attached to the brigade diary. On 9 April seven brigade members visited the Archenholder observatory together after work and decorated the brigade diary with photos from the catalogue. On 12 May eleven members went bowling at the Restaurant Freundschaft on the shore of the Müggelsee. An order for a large table had been placed two months in advance. As can be read in the form filled out to book the tables, there was a menu consisting of 'Celestine' beef tea, filled pork rolls in breadcrumbs and ice cream with fruit and almond cream. On 6 June six women went to the theatre together with their families and friends, totalling twenty people in all. The delegate for culture of the brigade had already ordered tickets three months in advance in the Metropol Theatre for the *Graf von Luxemburg*. Yet, due to a change in the performance schedule, they had to go and see the operetta *Polenblut* instead. On 15 September four women, who received distinctions from the trade union directors for their work in the enterprise trade union committee, were allowed to fly to Budapest for a day. They went shopping and tried to see as much as possible of the city. Boarding cards, safety instructions and cabin information were all kept for the brigade diary.

On 29 September ten members met up with their children shortly before work finished at 4 p.m. on the outskirts of Müggelheim for an autumn hike together, led by an employee from the department of culture of Köpenick. He told them stories and legends about the origin of the Müggel Mountains, about giants and about the beautiful king's daughter at the bottom of Teufelssee. On 8 October the brigade visited Brecht's former house in Chausseestraße and afterwards went for a meal together in the restaurant Brechtkeller. Eight pages of the brigade diary were later filled with photos of Brecht's house and a report about Brecht and Weigel's lives. On 31 October five brigade members went bowling again in the restaurant

Freundschaft. We learn from the brigade diary that the champion turned out to be the only man in the brigade. On 16 December in the guesthouses L. and B. in Müggelheim the brigade organized a fondue meal to celebrate the end of the year. The fondue had already been ordered six weeks earlier for twelve people and the meat was in such abundance that the women filled a plastic bag with the leftover meat and used it to cook up a substantial goulash soup two days later in the enterprise.

The women in the brigade obviously enjoyed having a good time together away from their husbands and children – an enjoyment, moreover, they could legitimize officially as part of socialist competition. Among the process planning brigade at Stanex, which consisted mainly of men, the social dimension played a considerably smaller role. Yet, a day of hiking together was organized almost every year and was also carefully planned in advance. The brigade sometimes also went bowling together. In the 1970s cultural events, like visits to the theatre and exhibitions, were still commented upon, by the 1980s they no longer figured in the diary. International Women's Day was reported every year, even though the men often found it difficult to celebrate the day appropriately. On 8 March 1973 the entry in the brigade diary reported: 'One of the greatest weaknesses we have as men, forgetfulness, obliged us to overcome unexpected difficulties.' Grabher commented:

> 'We were to some degree lucky that we only had a few women in our brigade. So we could always organize something pretty fast. There were of course other departments, like in the materials management or in the sales department or in accounts, where you had two or three men and then perhaps ten or twelve women. The lads had quite a rush to get something for the women. In those departments the women often got at least some coffee and cake and then maybe something out of the brigade money chest, like some sort of little gift or a small bottle of perfume.
>
> Once, I still remember, I heard them all whispering: 'Our lads haven't prepared anything for us!' Because they always got wind of it. But up there in the TV tower, in that rotating restaurant, I had booked a table for five: for three women, for a guy out of the same department and for myself. I said something like this: 'Keep quiet! Don't say a word before the day, not until 8 March!' I had this table booked for 10 a.m.! Right after 9 we put on our coats and said: 'Come on, get your coats on! We've got something planned for you!' And that's exactly what we did … a complete success! They didn't even get a bunch of flowers in the morning or anything else! Nobody did anything like that, as if we had forgotten. And were they annoyed! And then at 9, it was just time for breakfast, I said, 'No, don't eat anything now.' – 'Well, why not?' I said: 'Come on, put your gear on, we're off …' (Grabher, process planning, Stanex, 28 May 1991)

'That's life', said Grabher, satisfied years later at how he had made his colleagues happy. Even if the occasion for this was International Day for women's struggle, this gallant excursion to the TV tower had nothing to do with struggling. On the contrary! The women who had been taken out for the day placed themselves in the care of the men and let themselves be spoilt. The arrangements for this special occasion had completely detached the event from the original meaning of Women's Day, allowing the men to play the classic role of gentlemen.

The commitment to 'live the socialist way' encompassed being together socially, but also participation in political events, demonstrations and in civil defence. From the mid-1980s the process planning brigade at Stanex no longer mentioned any participation in demonstrations. The purely ideological statements in the brigade diaries became noticeably less frequent. The last political protest resolution which the brigade members wrote up and attached came from 1984 and was directed against the NATO mid-range missile deployment in Western Europe. In previous years resolutions entered the diary against the 'USA bombing terror in Vietnam' (1973), against 'Pinochet's putsch in Chile' (1973) and against 'Israel's brutal attack in the Lebanon' (1982).

From the mid-1980s there was a leap in the number of members in the Society for German–Soviet Friendship. The brigade declared itself ready to participate in a special competition for the title: 'Collective for German–Soviet Friendship'. They wrote in the competition statement that they had created a wall newspaper about the Soviet Union, organized a slide show and discussed *Sputnik*, the Soviet youth magazine, whose German edition was withdrawn in November 1988 from the GDR mailing list. A book reading of a modern Soviet author, which they wanted to attend, was cancelled too, because of the tensions between the Soviet Union and the DDR, and the general assembly of the Society for German–Soviet Friendship was postponed. The old slogan 'learn from the Soviet Union, learn to be victorious', acquired a new dimension as Gorbachev's reforms came into being. The slogan was used with new fervour in the enterprise.

Some members of the brigade continued to participate in civil defence exercises as late as 1989. Grabher himself told me with enthusiasm about his own service in the civil defence for over twenty years: how they had simulated emergencies in times of crisis, lowered 'victims' by cable on a stretcher from the sixth floor and developed strategies for the eventuality of a jumbo jet crashing in a residential area. In his opinion, from the beginning of the 1980s the civil defence had lost much of its military character: instead of simulations of nuclear attacks, planned with military precision, only civil catastrophes were simulated.

Many brigades became the official patrons of school classes, which they accompanied for several years. In the contract that the Blue Light collective[2] from Stanex agreed with class 2b of the prestigious sports school on 18 October 1974, the class committed itself to inviting members of the brigade to special events, like Pioneer's birthday, Children's Day on 1 June and the handing out of school reports. They also composed wall newspapers for the brigade. The teacher informed the patron brigade officially about the performance of the class and it became an external authority to which the pupils had to prove themselves. The brigade invited the class regularly into the enterprise. In 1974 they went together with the children on an excursion to the Niederfinow ship hoist. Members of the brigade took part in parent – teacher evenings and gave special awards to the brightest or most hard-working pupils when the reports were handed out.

> Earlier on it was thought that our strategies would 'guide the children towards production and the working class'. That was all rubbish and none of it ever came about! But I always felt that a certain understanding stays with such kids. In fact they always developed a trusting relationship. When they got their reports, we celebrated handing out the reports and I put effort into this too. In agreement with the teacher, we put a few pennies from the brigade treasury into a kitty and bought a few books for the best pupil, well, not just for the best one, but also for those who were a bit weaker. Anyone who made a real effort to improve from mark 3[3] to mark 2 or from 4 to 3, they often had to make a greater effort than someone who simply manages a mark 1 and then sometimes gets 2 and then 1 again – it's them who always get the praise, so those who really try hard and swot up at home, so that they can get a higher mark for once, they don't get the recognition they deserve. We always took that into account a bit as well. The pupils would be thrilled then. (Grabher, process planning, Stanex, 7 May 1991)

The men from the Blue Light brigade enjoyed showing the children the complicated machinery and demonstrating their skills to them. As official patrons of the school class, they were to set a good example and they enjoyed it to the extent that they thought about not only rewarding the best pupils but also those who had made a considerable effort. They tried to counteract the tough discipline and competitiveness of the sports-school, through encouraging those children who were not destined to become the best. Grabher stressed the direct friendly relationship they developed with the children and denied the supposedly political function of this friendship.

The obligations taken on to 'work the socialist way' show implicitly in their detail the weak points in the organisation of work in the brigade, and in the cooperation between departments in the enterprise. The brigade committed itself to accomplishing tasks that were necessary, but which the plan did not provide for. They acquired materials needed through informal channels and helped out in other areas of the enterprises, although they were often not qualified for the tasks.

When the fulfilment of the competition was assessed at the end of the year, it appeared that the promises to improve the cooperation within the enterprise had in fact been difficult to accomplish. The members of the Blue Light collective admitted that they did not draw up the monthly operative plans by the 20th of that month as promised. The blame for not being able to draw up the plans in time was put onto the materials management, which had failed to obtain the production materials by the agreed deadline. As production had to continue, process planning had to acquire the necessary materials in informal ways. This was a circumstance which was elegantly phrased in the accounts as 'special socialist work efforts in material management', meaning informally procuring materials, or as 'providing documentation', meaning writing lists of missing parts.

Members of process planning helped in production, prefabrication and 'organisational jobs' of other departments. In 1973 the collective worked together on a special programme for prefabrication, provided input to a government assignment outside of the plan for antenna production and carried out work on heavy loads. The engineers of process planning returned to production for qualified, as well as for unqualified work, manned machinery or hauled materials, filling gaps in the irregular course of production. They could then declare these uneconomical ways of working as exceptional work performances. The symbolic political signal they gave counted for more than their actual assistance in the production. By returning to work the machinery, they temporarily did away with the distinction between physical and mental work. By taking on the function of gap-fillers, they underlined the shared responsibility for production under the planned economy and legitimized planning itself.

The brigade diaries provide a detailed perspective into the life and work cycles of a brigade. When interpreting the information supplied by the diaries, it must be taken into account, however, that they were only kept in order to participate in the socialist competition and to be entitled to the bonuses that could be gained. Spontaneous voicing of opinion was not acceptable in the brigade diaries, which were read and assessed at the end of the year by the enterprise director, the party secretary and the head of the enterprise trade union. Since bowling evenings, dance events, visits to the theatre, even weddings and visiting the sick could all be accounted for as part of socialist competition, they ceased to be exclusively part of the private sphere. Reports about social gatherings of the brigade often reveal ambivalence, distance and a certain uneasiness. This becomes particularly clear in a report about a dance event held on 8 March 1974, in which the author only uses passive sentence constructions (was carried out, was created, were entertained, were treated etc.). Here is an extract:

> We were well entertained by the musicians and dancers.
> We were treated to a stand-up banquet after this culture programme.
> This stand-up banquet was seen as a welcome change.
> After this stand-up banquet the hips were swung into action for some dancing.
> The evening was called to a close around 2 a.m.

'Pleasure became a duty', as a Stanex employee put it. Accordingly, the Blue Light collective wrote in the competition accounts at the end of 1980: 'Living the socialist way still has some shortcomings.' Lively reports were rather the exception, such as the following one from an employee of process planning, who in November 1973 rode his bicycle through East Berlin to get hold of some sticky tape for bicycle handlebars, which was necessary for production.

> On that particular Tuesday the 13th of November, we had dreadful weather, but the situation was such that, it became necessary to get the tape for the handlebars.
> Colleague Conrad organised some money and Colleague Baumann then rode his bike on the hunt for the tape.
> After he had unsuccessfully 'raced past' nine bicycle shops, he didn't give up – on no account did he want to return without the handlebar tape.
> Filled with this thought, he fell upon the idea of getting glazing tape for windows instead. He was lucky enough to find some of this in a paint shop in Münzstraße.
> He returned jubilant and exhausted. The LS 3-production could go back on line.

The report is unusual because it openly allows the production problems of the planned economy to surface: production stoppage due to lack of supplies, improvisation using improper materials in the production process, empty shops, resorting to the spontaneous initiatives of the employees. The mask of the socialist performance is removed and the unvarnished reality is revealed.

Covertly …

When I interviewed the workers and employees in 1990 and 1991 about the socialist competition, they had few positive remarks to make. Only Grabher explained to me that he had felt the cultural events now and then to be beneficial and to be a welcome opportunity to talk with his colleagues about things other than their everyday work. He commented about the brigade report of the visit to a photo exhibition for the fiftieth birthday of the USSR on 12 February 1973:

> Some perhaps got some inspiration out of it, others didn't. To give intellectual inspiration was anyway not what was intended with this exhibition. But there was indirectly perhaps something that was a bit useful, when you ignored this political pomp for a minute, when you regarded that as a theatre backdrop and really observed the people [in the photos; B.M], when you made the effort to advance towards the core, the centre of it all. And also one would never have gone to such an exhibition on one's own accord. (Grabher, process planning, Stanex, 28 May 1991)

Referring to certain positive aspects in the practice of socialist competition, Grabher justified his long-term function as brigade leader. He attempted to separate 'the human element' from the 'political pomp' and to place it in the background as a 'theatre backdrop'. Through the political performance he wanted to advance towards the 'real core' and to discover a meaning in these not quite voluntary activities they did together.

Nevertheless, most of the retrospective comments his colleagues made regarding socialist competition trivialized political life in the enterprise. The interviewees vividly depicted how they transformed the function of the cultural events and of the compulsory political events into moments of social get-together. During the 1 May demonstrations a few brigades went into the nearest side street to the meeting place to go and have a beer. On a visit to the Soviet memorial in Treptower Park most of the colleagues met up afterwards in the pub. As one worker defined it, you enjoyed the small pleasures of life using the competition as a pretext. Trivializing the political life of the enterprise might appear as a form of resistance to the political paternalism of the state (Scott 1990). But the analysis of the interview material shows that no bipolarity can be assumed, no either/or of obedience or resistance towards ideological dominance. There were forms of keeping a distance, which are not meant as forms of resistance or resisting, even if they occasionally appeared like this. The fact that being together was a duty in some brigades disrupted rather than encouraged communal ways of life.

> The employees had been closer together earlier on, before we started fighting for the cause of the famous socialist collective – 'fight' is not to be taken literally. When we were under no obligation to be together in a work collective, we also did things as a group, even going on trips abroad with one another. But when this trend was introduced, and was made obligatory by the state – logically nobody wanted to become this famous martyr and say: 'I'll take up the flag and run head first in protest against this change!' – instead everyone withdrew a little and the idea of this big pleasant togetherness clearly received a setback. Everyone withdrew a bit more into their shell, so that in the following years the things we did together became less and less and, well, strictly speaking we just put up a show for everyone else. (Mahler, project planning Stanex, 2 May 1991)

The obligation to stay together destroyed the joy of being together and the brigade members perfected the art of putting on a performance. Even figures for balancing the competition accounts were invented in such a way that they seemed plausible – not too high, and not too low. The lively communal life disappeared to the extent that it turned into an obligation. Socialist competition forced the participants into pretending to be abiding by the political and ideological system. On a small scale, accounting for the competition emulated what was being practised on a large scale and daily in socialist planning. A fantasy was created, which helped those inside the system to carry on their lives in peace. In retrospect, many of the interviewees argued that the socialist competition had been 'insane'. They had been able to see right through it, but had to participate in it because the system in its totality forced them to do so.

> You really did look for this famous golden path that went right down the middle, following this trend of 'don't stand out, don't cause any bother', in order to be left in peace. One tried not to be driven to the front by being pointed out: 'as the best one'. It is not at all pleasant if you're sort of in the 'limelight'. Because in that position you see that the others think: 'Look how stupid they are!' You also had to watch out that you weren't passed over too much and that you weren't right at the back. That was when you hit the right balance. It would be alright then. (Mahler, project planning, Stanex, 2 May 1991)

The resignation Mahler felt at the time rings clear in his words, which describe how he guided his employees along this path down the middle in his capacity as head of the department and of the brigade. The brigade faced a further difficulty in that the director of the enterprise, Dr Schöpf, was formally a member of the brigade, although he did not participate in their activities. As a devoted party member, he valued highly the fact that 'his' brigade came off well in the competition and he personally checked the brigade diary and wall newspapers. For that reason Mahler had to prevent his brigade from being looked on negatively by the director, while preventing colleagues in other departments from seeing his staff as 'the stupid ones' or as particularly eager communists.

Yet even brigade members in enterprises like Taghell, which were deemed politically lax, were still full of rage about the rituals of socialist competition years after the Wall fell:

> Well, yeah, we did a few things, certainly. We went on different trips. In principle, the competition and everything was nothing other than a big lie. The whole lot was a mess, a load of rubbish … they all filled up each other's pockets. When you saw the thing, when it was up and running. The competition, what a load of nonsense it was, sometimes it was totally amazing, really. (Friedemann, fitter, Taghell, 29 July 1991)

Two years after the Wall fell, Friedemann spoke contemptuously of the competition, as 'a big lie', full of 'falsehood and deception', in which 'each and everyone' had a part. In his account, he appears to be merely an involuntary spectator himself. As a fitter he was not responsible for the leadership of the socialist competition, but part of his bonus depended on it. He 'filled his pockets' too. The central problem of the socialist competition rests in this contradiction. The frustration Friedemann still feels today is partly frustration with himself – frustration that he joined in.

In all three enterprises it is particularly striking that the most perfect wall newspapers and brigade diaries were presented precisely by those foremen and heads of department who were least committed to socialism. In so doing, they won for their brigade the certainty of being spared ideological and political issues and moreover of profiting from the different bonuses, which were available to those who had been successful in socialist competition. Being distinguished as the best worker or the best collective aroused ambivalent feelings. Even the most critical of workers swelled with pride if, after years of work, they were distinguished as an 'activist of socialist work' or received the service order of the fatherland in gold, like the skilled worker Ruland at Stanex.

Well, there was a precise profile for the sort of guy they were looking for:
Not belonging to the party, if possible lots of children, active in the trade
union, thirty years service for the enterprise, well, an honourable citizen,
so to speak, and then they would all be put on a list and of the two
thousand employees or so, half would be crossed off after the first
selection straight away. And that's how it would go on until one or two
were left and then it really isn't his fault if he gets the order. I'm telling you
that's how it was because when he got the reward many would reproach
him for that. His efforts for the trade unions, he made without any sort
of reward in mind and he never strove for, let's say positions, some
position in the head office, it was quite the reverse, he always tried simply
to continue working, something that many others didn't. (Schuster,
skilled worker, Stanex, 22 April 1991)

Having being awarded the order, Ruland was exposed to the mockery of
the other workers, who were also envious. He was made fun of as being
loyal to the party line and secretly a full-blooded communist, but there
were questions raised too, asking 'why him then? He's no better than
me'. Since it was important for Ruland to continue working in the
enterprise as a normal member of the collective, he tried to make as little
commotion as possible about his order and did not claim any of the
privileges that would have been available to him.

In contrast, the head of department, Voigt, at Stanex enjoyed being
awarded an order and wanted to show it off in front of his comrades and
co-workers. The skilled worker, Fröhlich, who was not keen on such
festivities, commented:

I know at some point he had been awarded an order. It was 'esteemed
metal worker of the people' or something like that. Then for the whole
day all the comrades were invited up into the breakfast room, where they
were generously treated to drink and many toasts went to this order that
had been bestowed. As for us, we could clock off work half an hour or an
hour early. Then we had to go up and sit down, where there were servings
of sandwiches along with beer and fizzy drinks. And up there he had
everyone admiring his order. I could not stand it. He passed this order
around and everyone was allowed to touch it. Well, it was awful! The only
thing I took away with me was a piece of cress for our guinea pig.
(Fröhlich, Stanex, 16 June 1991)

Fröhlich vividly depicted the party comrades, who on the occasion of
awarding the order to the 'esteemed metal worker of the people'
celebrated themselves and had a good few drinks. He snubbed the
spectacle, scorning the food and putting those celebrating and his guinea
pig on the same level.

In enterprises like Stanex, where ideological purity was valued highly, the brigades created wall newspapers every month along the lines of the large themes promoted by official party organs. The area manager evaluated them together with the enterprise group leader of the trade union and the head of department, and granted points, which counted towards socialist competition. Wall newspapers got positive marks if they neither stood out with unpleasant political statements, nor were an exact copy of the news-sheet from the previous year. The fiction had to be sustained that the wall newspapers were a spontaneous collective view of the workers, in natural agreement with the political directives of the party. A particular individual view, joke or humour, above all if it was profound and ambiguous, was unwelcome. The following poem that Fröhlich wrote in 1982 to celebrate Christmas as a feast of contemplation had to be removed at the orders of the head of production:

Contemplation (1982)

Ist es wirklich schon so weit	Is it really time already?
Ist das Jahr verschlungen?	Has the year gone by?
Ist es in uns tief und breit	Has it really reached into us
Richtig eingedrungen?	Deep and wide?
Schlug die Welle unsrer Zeit	Did the wave of our times
Manchmal in die Brandung?	Sometimes break upon the crest?
Folgt dem Flug der uns geweiht	Did the flight we embarked upon
Eine gute Landung?	Land safely?
Haben wir genug versucht	Have we also tried hard enough
Menschen auch zu bleiben?	To remain people?
Oder wurde nur gebucht	Or have we only registered
Durch den Sog der Zeiten?	In the vortex of time?
Weihnachten steht vor der Tür	Christmas has come once again
Zeit um nachzudenken,	Time to reflect,
Ob wir unsern Fuß und Geist	Whether our feet and mind
Immer richtig lenken.	Still steer us in the right direction.

Horst Froberg, (alias Fröhlich, 1982)

The declaration board remained empty for a week. Yet, when the director's tour of the boards was imminent at the year's end that established the department's position in the competition, the worker responsible put together a new wall newspaper, so as not to lose any money from the bonus. Almost ten years later in 1990, the memory of this disgrace still annoyed the workers.

In the department for project planning elements of wall newspapers were kept in folders and put up time and again. The two employees who put the newspaper together every month had fun in repeatedly using the same photos, articles, and commentaries without the area manager ever noticing. They stored emblems, portraits of political leaders and ideological propaganda texts, altered the names of the authors, carefully expunged any mention of a date. Whenever they wished to use a wall newspaper article again, they carefully put the drawing pins through the existing holes from the previous year.

> Schadorf: 'We had become so proficient at this that whenever articles had to be written up for certain festivities or state celebrations, we compiled them without any dates or precise details of the actual time the articles were created. That way we could hang them up again and again every year. We only had to find the same holes and use them for the pins when putting up the articles.'
>
> Born: 'We were careful to have only one pinhole and to work with extremely fine pins because Schöpf noticed things like a pinhole: 'This has already been used three times and new holes had been put in.' So we got to know all the little tricks … in the end we even did sketches of how it all had been put up before. The article for the 1st of May hung up in the corner the previous year. Then we said to hide that away better, we'd have to change the pieces around some more. We had really got quite adept at this because we started having our mischievous fun in outwitting the others too. It is in fact quite sad when we think about it afterwards. But this meant most of all we were left in peace. And if you were to ask the others, they weren't even aware of it.' (Born and Schadorf, project planning, Stanex, 4 February 1991)

Both of the engineers put much effort into putting on a perfect performance of the official ideology. Their cynical distance from their own performance rested most of all in their refusal to put something of their 'own' onto the board, like a text they wrote themselves or an article about a subject that really interested them. In their effort to construct an apparently 'up-to-date' newspaper out of standard elements in the shortest time possible, which even Dr. Schöpf accepted, they proved the banality of the ideology to be reproduced, as well as its disconnection from reality. Furthermore, they also confirmed the inanity of their superiors, who were unable to discern an 'up-to-date' wall newspaper from one of the engineers' fabrications. As they did not share their fun with their colleagues, their scheming could not take the form of a common conspiracy against the ideological paternalism of the state. In retrospect, a year after the fall of the Wall, it made them feel sad to think how much energy they had invested in putting on their performance.

With the collapse of the socialist system, the ideological compulsions they had lived under seemed almost unreal, and their hidden resistance, which had them put drawing pins in the same holes over and over again, seemed trivial. Their cheekiness had given both of the engineers a feeling of moral and intellectual superiority over their bosses and the ideological system, but at the same time it made them contribute to its effectiveness. As Scott expressed it: 'If rituals of subordination are not convincing in the sense of gaining the consent of subordinates to the terms of their subordination, they are convincing in other ways. They are, for example, a means of demonstrating that, like it or not, a given system of domination is stable, effective and here to stay' (Scott 1990: 66). By successfully achieving the perfect imitation, the two engineers could indeed deceive their superior, Dr Schöpf, who wanted to force his subordinates to degrade themselves to the point of formulating genuine statements praising the system, despite the fact they did not support it. However, from an external perspective their wall newspapers were not unlike any others. This private way of keeping a cynical distance from power relationships was in terms of behaviour, hardly any different from voluntary compliance. It had the effect of strengthening power relationships rather than undermining them.

Their superior, head of department Mahler, also spoke with regret about the 'infantilisation' his brigade was exposed to through the socialist competition:

> Strictly speaking adults allow themselves to be degraded and perform such baloney. That's exactly what is humiliating about it. (Mahler, project planning engineer, Stanex, 2 May 1991)

Not every brigade member made a contribution to the brigade diary, but each benefited from the bonus that the brigade obtained for a successful competition. Whoever strayed openly out of line and disturbed the performance of the brigade, disadvantaged the group as a whole. Even joining in without conviction supported and legitimized the official political-ideological line.

The wall newspapers and brigade diaries fulfilled the function of gestures of subjugation within the logic of the system of the planned economy. As the workers, seemingly at their own initiative, discussed the main themes of socialist economy, criticized themselves and promised greater performance, they formally legitimized the system and their role in it. The complex system of balancing the competition went along with individual and collective shifting of blame. As individuals and collectives admitted to their incapacity to fulfil the targets they had set for themselves, they indirectly assumed responsibility for the inadequacies of the system.

It is not easy to estimate after the fact how much of socialist competition was indeed nothing more than a pragmatic performance, which was passed on to those in power so that 'we-down-below can have our peace', and how much was indeed a result of a deep-seated belief in the system. The ambivalence of socialist competition arose from the amalgamation of the political and the private. The competition celebrated personal relationships between brigade members as part of socialist life and deprived them of their spontaneity. On the other hand, the political obligations of the brigade contained personal elements or were redefined as social events. This association made it more difficult for the brigade members to reject the competition straight out and to withdraw from the communal events and activities for political reasons. This distinct atmosphere emerging from personal warmth and political control, characterized the GDR more than any other socialist country.

Notes

1. *Gedanken* (1984)
    Was ist es nur, was uns so quält
    Beim Grübeln übers Leben
    Wer ist es, der die Richtung stellt
    In die wir uns begeben?

    Warum ist mancher Kopf so leer
    Ein anderer quillt über
    Der eine stellt sich schrecklich quer
    Der andere lächelt lieber

    Wie heißt der Motor ringsumher
    Wo ist das Ziel zum Tanken
    Der Eine, der fällt sofort um
    Ein Anderer muß erst wanken

    Auf Erden, da ist alles rund
    Drum eckt man ständig an
    Und wer hier nicht ins Grübeln kommt
    Der macht es nebenan.
2. The Blue Light collective takes its name from the flashing blue lights on police and ambulance cars. Responsible for the logistics of production, they played the role of troubleshooters.
3. In the GDR school system, mark 1 was the best, mark 5 the worst mark.

Chapter 4

Party Rule in the Enterprise

Preparation of the conference room on 31 October 1988 from 2 p.m. onwards:

- Executive committee for 13 comrades, tables to be covered with red cloth, potted plants.
- Speaker's podium to be draped with a red flag and the SED emblem
- Audio equipment with 2 microphones for the speakers podium and executive committee.
- Flag poles with the BPO [party organization of the enterprise] flag, GDR flag, AKL flags.
- Provide voting urn, transportable voting urn.
- Seating for approx. 185 comrades, tables to be covered with white cloth.
- The current slogan is to be used.
- 1 table for the editorial commission.
- Responsible for arrangements APO 3 [*Abteilungsparteiorganisation* – department party organization] and APO 5 each with 3 comrades respectively.
- Rearrangement of everything back into its place in the restaurant is to be carried out by APO 2 and 4 each with 3 comrades respectively.
- Record player, records with workers' songs and revolutionary songs.

Extract from the *Planning steps for preparing the election meeting of the party organization of the enterprise Stanex*, 19 September 1988

The political control of the economy by the party and the plan, which in turn received its direction from the central committee of the SED (Socialist Unity Party of Germany), constituted the key difference between market and planned economy. The 'ideological work' undertaken by the party organization of the enterprise might appear as a completely superfluous task when seen from the perspective of the market economy. However, production in the planned economy was not separate from the state, and non-economic considerations, such as political control, social redistribution, and military strategies, were inseparably bound with productive aims. The ideological monopoly of the Communist Party stipulated state property and state monopoly over the economy, which in turn required bureaucratic control and coordination and the abolition of the market (Kornai 1992: 360). The party groups in the enterprise were the lines of transmission for decisions made at the head of the party and they had the task of reproducing and legitimizing the decisions of the central committee at the level of the enterprise.

The administrative hierarchy in the enterprise was doubled by the party hierarchy. The director of the enterprise who was supposed to implement the political directives had a party secretary at his side without any practical obligations. However, the influence of the party was not equally strong in all enterprises. At Taghell the party group was small and not very active, and despite the poor working conditions many employees had chosen to work for the enterprise precisely because of the lack of party control. In contrast, at Stanex, where some departments also produced parts for weapons, the party was omnipresent and in the 1980s, the department where I did fieldwork, turned into a scene of fierce political confrontation. In both divisions of Stanex in Berlin were 1,200 employees; 175 of them were members of the party organization of the enterprise, which comprised six department party organizations meeting every fortnight and these were respectively subdivided into party groups. The department of assembly automation consisted of 100 employees and had its own party group with fourteen members. The design engineer Karst was its chairman in 1988.

For the employees at Stanex the debates about their commitment for or against the party had emotional significance for years after the unification. They reflected upon socialist values and norms, which at the time of the planned economy was fiercely contested terrain among party followers and opponents, as well as inside the party itself among reformist and orthodox communists. Criticism of the ruling system only circulated in secret, although (or perhaps precisely because) it mainly moved within the framework of socialist ideology. Criticism was not principally directed against socialist ideas, but against the rigid and dogmatic interpretation they received from the party. In this chapter, I will analyse how the party leadership's moral claim to absoluteness became an instrument of power and turned into a guiding principle for the conduct of those in power.

Party and Career

The SED laid claim to knowing the only correct path leading to social and economic development. The party members' position of power was legitimized through this claim to absoluteness, but was at the same time restricted and monitored by the party hierarchy. Party membership meant taking part in the control of an instrument of government and at the same time being required to show absolute loyalty. The party could only really exert power over comrades. It was not until the economic directors had become party members that they could be committed to the political line of the party. Political conformity was as important here as professional competence.

Also, the economic directors who had become party members without the necessary political conviction, and joined above all to further their careers, had to represent the official ideology externally and prove their political reliability, for example, by keeping a close eye on the socialist competition in their enterprises and by accounting for their political and economic performance before the superior party bodies. Several non-members told me that most of the people who joined up out of opportunism and for better career prospects soon after proved themselves to be faithful to the party line even when talking in private. Scheuch, a former design engineer at Stanex, told me of how he ceased to be friends with Mayer, the later general director of the enterprise, when Mayer joined the party and then tried to 'convert' him:

> As far as I was concerned, when Mayer joined the party sometime in the seventies he had already as good as vanished. I got to know lots of people in my time who took this step for some career reasons and then all of a sudden tried to convert their old mates and would say: 'Look. Now I know what the right thing is for all of us and you all should know too.' (Scheuch, Stanex, 24 June 1991)

When Mayer took on management roles and therefore publicly and ritualistically began to propagate ideas, which were not originally his own but from the arsenal of the SED, he may have genuinely experienced these ideas and emotions himself while l performing them. In the longterm, cognitive and emotional dissonance led to a distortion of the 'boundary between the perception of a role played out from a distance and the perception of the self' (Rottenburg 1992: 244). In many ways, joining the party meant a break with previous social relationships. This had a deeper impact than, for example, the ban upon sustaining connections with relatives in West Germany. Party members spent less and less time with acquaintances who were critical of the regime and with those outside of the party organization. To the new member's own surprise, he/she was often forced to account for his/her choice of party membership to his/her long-standing friends or was outcast from his/her old social circles. This facilitated efforts to integrate the new member into party structures and to keep him/her under observation. The design engineer Kater told me that many comrades in the enterprise indeed preferred to keep the 'party manual hidden away in their pocket' rather than have a political debate with their colleagues. Anyone who publicly represented the party line, was performing 'ideological work'.

> Ideological work was mainly a term to describe how you should simply represent your views just as you normally would. There were lots of comrades – precisely these guys who only went with the flow – who generally preferred to avoid such discussions. Yeah, they were the people who would keep quiet or just walk past. Actually, ideological work only meant that the comrades who had signed up as members, had thus declared themselves to be supportive of an idea, which they were really then supposed to represent, not only to have the party book in their pocket and then if possible not to show it to anybody … (Kater, design engineer, Stanex, 25 April 1991)

The ambiguity for many comrades did not necessarily reside in the fact that their own point of view did not agree one hundred percent with that of the party, but that they could not be certain which was the right stand to assume. They wanted clarity so that they could then adopt it as their own and defend it, without getting into any danger of straying too far from the official party line. Career-conscious comrades preferred to keep quiet rather than open themselves to attack. Whoever defended their own individual stance exposed themselves to criticism from other comrades and attacks from non-members.

It was difficult for a party member to switch enterprises and be integrated into a new work group. Frequently his colleagues regarded him as a spy working for the enterprise management, one who was supposed to listen in to their conversations and to motivate them to better performances at work. The skilled worker Schuster described his initial experiences as a comrade in production at Stanex.

> At the time there were two comrades in the whole contingent. In a word, they weren't at all popular. An outsider, alright he has to try and fit in anyway. But then to be an outsider and a comrade, that's to say, a party

member. That was unpleasant enough. They didn't know me at all. First of all they looked down on me like I was some fool from the party. What they didn't know was that, despite being in the party, I had a view that was completely different from that of the general party line. (Schuster, skilled worker, Stanex, 22 April 1991)

Schuster was eighteen when he joined the party, as he put it, 'out of youthful enthusiasm' as a 'political activist' – a step that shocked even his father, who was a full-time functionary and member of the diplomatic corps. 'I found it normal, in this state, I also found it correct and good. My parents had led their lives up to that point with the belief that socialism actually had a future. As a young man you don't have much to compare your life with.' These days he thinks his membership of the party did more to prevent his professional progress than encourage it. Although the enterprise wanted to make it possible for him to begin studying to become an engineer, Schuster preferred not go ahead with this because it would have required him to do more 'political work' in the enterprise.

Careers in socialism always experienced a rift at the point where the persons to be promoted refused to join the party (Lambrecht 1989; Niethammer, von Plato and Wierling 1991). Therefore, due to refusal of party membership there were frequently more competent managers at the second level, while the highest positions – those of the directors – were occupied by incompetent people. These circumstances are exemplified by the manner in which the career of the engineer Grabher was cut short at Stanex .

Grabher's promotion to director of production at Stanex in 1975 demanded obligatory membership in the party. Grabher vividly described how party representatives from the enterprise and district office spoke to him for hours on end in their attempts to convince him to join the party:

At the time I had to endure talks, which lasted for ages: perhaps three, four or even five hours. It would start sometime in the afternoon around three with a cup of coffee and a cognac and then go on into the evening until about seven, ranging from friendly chatter to very clear and distinct demands. … In many conversations they desperately tried to tell me: 'Well, through your work and through your performance in the workplace and your views … You are in fact standing right behind the party. All you have to do, is simply sign up and join the party. After all, you are a socialist.' (Grabher, process planning department, Stanex, 7 May 1991)

Nevertheless, Grabher did not want to join. He justified his refusal to join by explaining that for him it would have meant 'surrendering the last remaining tiny fraction of his personal freedom'. But he did not mention this reason to the comrades. When talking to them, he used the argument that he did not want to give up the contact he still had with his brothers and sisters living in West Germany.

> So then I said: 'What about if I apply to become a comrade and director of production, but then I want to visit my brother because of this and that? I need the necessary approval from you!' People needed a permit from the enterprise, otherwise the police wouldn't even consider your visa application. They needed authentication from the enterprise. I said: 'In that case, you will have to convince me until I say, oh well I don't want to visit my brother at all any more.' (Grabher, process planning department, Stanex, 7 May 1991)

By using his private life as an excuse, Grabher had sidestepped the intensive efforts by the party to entice him. In so doing he could avoid the considerably more unpleasant political-ideological discussions, where he would have had hardly any chance to oppose them without publicly displaying opposition to real existing socialism. All the same, as he had put family over party interests, he disqualified himself from being a socialist director. After this conversation, Grabher was crossed off the list of potential party candidates. He stayed on as head of the department, and even took charge of a new Stanex department in 1978, but he did not rise any higher than this position.

For many of his colleagues in design engineering and in the process planning department, it was not worth striving for promotion to become head of the department or to gain a seat on the board of directors, because this cut them off from their skilled work and assigned them political 'social' tasks, which they did not value very highly. Karst, an orthodox party member and design engineer at Stanex, explained:

> We were able to work independently. We had our own work groups, so that we could stay within our area of expertise and didn't just regurgitate the plan of the enterprise. Anyone who had a managerial role did some job or other, but didn't do any skilled work. The group leader still did some of the skilled work, but if you go any higher up the scale than that, then you did not do any skilled work at all! (Karst, design engineer, Stanex, 7 February 1991)

Party Control

Being a member of the ruling party was neither a hereditary status like being a noble in an aristocratic society, nor was it a prerogative acquired like becoming a wealthy person in a capitalist society. Party members had to repeatedly prove themselves and this coupled with the mistrust they faced from their own comrades, kept them in a constant state of insecurity. Party members, even if they were critical of the party, could see themselves as important, because the party committees discussed social issues as if the future of humanity were in their hands. Furthermore, party members enjoyed certain privileges. These privileges, even if they were petty and seemingly insignificant, served to distinguish them from those around them. The members felt that, because they were part of the party power structure, they could bring about changes in society.

In fact, the party members were supposed to act as instruments in fulfilling a historical dynamic, which was established through marxist-leninist and empirical technical science (Meuschel 1992: 194). Discussions were permitted within the party, but the outcome had to be consistent with a view that had already been imposed from above, which the comrades were meant to represent with absolute resolve and tenacity. The party members were supposed to pretend decidedness and avoid decision[1] (Niethammer 1990: 259).

> Publicly we were supposed to represent the official party line – there was still this so-called democratic centralism – even if you didn't agree with it yourself. I'm under the impression that in the beginning, or let's say up until mid-1980 to 1985, this was hardly discussed or was perhaps never discussed beforehand either ... But afterwards discussions took place, whether allowed or not, and they became more heated because the problems became more pressing, although no conclusions were actually reached because any decision always had to come from above. Although we could discuss the decisions which were reached, they still always had the last word on any matter. This is because this democratic centralism stipulated an eligibility from the bottom up, but specified decision making from the top down. (Kater, design engineer, Stanex, 25 April 1991)

In their reports party members used their own ideologically coloured language, which was unlike normal language. Instead of representing a collection of common beliefs, the ideology had deteriorated into a catechistic recital of standard principles (Bendix 1974: 347). I could read these principles time and time again in the enterprise party newspaper and in reports put together by the party organisation of the enterprise, but in 1990 I no longer heard them mentioned in any of my conversations with the employees. Although these principles were supposed to be programmatic and future-oriented, they were in fact meaningless.

Since the fundamental underlying principles were never allowed to be placed under doubt, they each turned into dogma. These then provided the basis for further argument, depicted for example in the sentence: 'The key accomplishment of the working class is the revolutionary party.' It implies the working class achieved the power of the workers' party in the revolutionary upheaval and that power was thus wielded in their interest. This was arguably not the case in the postwar GDR, where the power of the party was established by the occupying force of the Soviet army. In contrast to their superiors, workers tended to join the party less and less and they frequently saw their interests as contrary to those of the party members.

Since all political discussions in the party organizations were based upon professions of belief, any argument was prevented from advancing beyond these foundational truths and became circular. Discussions revolved around absolute truths but could not convince those who did not accept these basic principles, and who thus had to be 'converted'.

Members were meant to represent the party view assertively and were not permitted to conceal the fact that they were party members. Therefore, although they were placed in a superior position to non-party members they were also separated from them. The party used this as a mechanism to sustain its absolute authority.

> It must prevent the individual member from keeping things to himself, and it must prevent him from hiding before his fellows all the opinions and actions, which would betray his identification with the party. If he kept things to himself, the party would run the danger of thoughts and potential actions subversive to its rule. If he could hide his identity as a party member, he could have a private life among his fellows, because they would not be put on their guard against him; but then he would be beyond the grasp of the party and, hence, a potential threat to its absolute authority. (Bendix 1974: 416)

The public opinions given by the comrades always had to conform with one another. Contradictory views, although they existed everywhere, were systematically suppressed during public declarations. Anything that was allowed to become public had to be in accordance with the official party line, which was put forward by the ZK (*Zentralkomitee der sozialistischen Einheitspartei* – Central Committee of the Socialist Unity Party of Germany) and confirmed at party conferences.

> It was always the case that the party – because they were all communists – were always right! And that fact managed to establish itself somewhere up here inside your head! And that's how this basically Stalinist attitude got deeper and deeper into your mind. It actually got worse and worse, instead of getting any better! Never during a public discussion in a party conference was criticism ever uttered! You should have listened to one of these party conferences! It's enough to make you put your hands over your ears! No mentally alert person could ever have listened to that! (Schuster, skilled worker, Stanex, 22 April 1991)

Even as late as 1989, opinions of party groups and trade union organisations were carefully checked for ideological purity. To illustrate this more clearly, the skilled worker and comrade Schuster showed me a paper he had written, in collaboration with a design engineer, a non-comrade, for a party conference. In the paper he pleaded for building more industrial automates in the GDR and stressed its importance for the national economy. The paper had been corrected and censored by five different people from various levels of the party hierarchy. Any tiny reference that might have revealed the poor technological standard of the GDR economy was cautiously eliminated and positively reformulated. For example, the following paragraph was crossed out:

> As you will know, improved automation of the assembly process has become a pressing necessity in order to guarantee the planned increases in labour productivity together with a simultaneous decrease in costs, an advance in the technical standard of the assembly processes and an improvement of working conditions.

This paragraph was replaced by:

> We began to consider the tasks of the 5th ZK conference of the SED and to draw conclusions from it. Comrade E. Honecker demands even more distinguished performances with high economic utility. This is first and foremost on our minds.

Although Schuster had kept to the official writing conventions and had referred to decisions made at party conferences, the formulations were trimmed down even further. When after the fifth phase of censorship all insufficiencies in technological development had been successfully rewritten, Schuster withdrew and decided not to deliver the paper.

> You have to see that in the whole GDR everyone took refuge in their own fantasies! So party functionaries only wanted reports of success, nobody wanted any more criticism. I always felt that we could only ever bring good news to our superiors – which means to the next highest committee up on the scale! And this is even if the so-called 'success' was just a pile of lies! But that's the way it was, I'm just not able to tell you why. (Schuster, skilled worker, Stanex, 22 April 1991)

Less important papers were subjected to the same mechanisms of ideological control. Employees at Hochinauf, who were in the BGL (*Betriebsgewerkschaftsleitung* – enterprise trade union committee), presented me with a draft written for the 1987 end-of-year report from the BGL. This report was written according to the same guidelines every year and described the work undertaken by the trade union. After three different people had edited the paper, every reference that might imply a mistake or inadequacy was carefully deleted. For example, the sentence: 'The quality of lunch needs improvement',[2] was reformulated to: 'The good quality of lunch, however, has still to be kept unchanged.'[3] Every little attempt was thwarted, whenever efforts were made to change or reform social or professional practice officially from below.

> The practice was actually steered in another direction by the party! So if you came along with a good idea, you had to go through the party, like a train approaching a set of points at a railway junction. You either got derailed from a 'sorting' track or they set you off in a direction that suited them! (Schuster, skilled worker, Stanex, 22 April 1991)

Although party members were exposed to permanent review and criticism in the party group, they were not permitted to let any criticism of other comrades or of the official party line ever get out. In some work contexts, this went as far as forbidding comrades from criticizing each other in front of non-comrades. The skilled worker Schuster described an argument with Karl Voigt, his head of department and comrade, to whom Schuster was superior in the party hierarchy, but inferior in the enterprise hierarchy:

> I openly confronted my head of department – we don't really get along. I wasn't allowed to do this because I was a comrade!
> You can ask Horst [Fröhlich; B.M.], he was there, and later on he couldn't stop laughing. ... well, I often criticized Karl openly! He tells me: 'You are forbidden to criticize me at all, you are a comrade after all!' So that means: as head of department he is above me in professional terms,

> but I was above him in the party because I had already been in the central party leadership for a good while. So I couldn't criticize him as a comrade head of department because he is part of my stables and he could do the greatest nonsense – I had to approve of it because he's a comrade! (Schuster, skilled worker, Stanex, 22 April 1991)

Instead of fending off the criticism from the subordinate employee by pointing out that, as head of the department, he was the boss and his instructions were to be followed, Voigt referred to their equal status as party members, which ruled out entering into an open argument. In reverse, Voigt, who lacked the professional competence to control the process of production effectively, attempted to discipline at least the workers who belonged to the SED and who were subject to party discipline. To set an example, he imposed draconian disciplinary action on them even for minor offences. During another conflict, Voigt tried to condemn comrade Schuster to severe disciplinary action as a result of a relatively small offence.

> I once infringed upon the rules by bringing my own private welding tool into the factory and trying it out. Karl [Voigt; B.M.] comes along, sees it and starts taking disciplinary action against me. Just like that! According to labour legislation in the DDR – I looked into it properly then for the first time – that wasn't right at all, but he put it through all the same! He wrote me a formal letter – I still have it, so I could find it somewhere – and instead of saying 'colleague Schuster', as in this case I am a colleague after all, it states 'comrade Schuster has violated the rules'! It made a big difference to me that he was no longer calling me to account as a colleague and did not position himself as head of the department, or colleague Voigt, instead he worded it so that comrade Voigt is calling comrade Schuster to account because he got him in the end! At last he had found something he could use to harass me. But in the end the action he tried against me did not go through. (Schuster, skilled worker, Stanex, 22 April 1991)

Schuster affirmed that he would have accepted disciplinary action against him as a 'normal employee' in accordance with labour legislation, but he did not agree with being called to account as a 'comrade' for such a mild misdemeanour. He opposed a disciplinary process, which would have subjected him to a humiliating interrogation by the enterprise party committee. He wanted to be judged according to law and not as a 'comrade' in line with moral political standards.

Part of the party's power resided in the influence of moral judgement. The individual was faced with the party's quasi-religious claim to absoluteness and was therefore required to behave properly, but was by definition regarded as fallible and blameworthy. 'Noble values' and a 'higher moral rhetoric' singled out 'most of the speakers in GDR society, whether of the intelligentsia or the opposition' (Meuschel 1992: 308). The party had known for a long time how to direct criticism of the practices in GDR society back onto the flawed individuals, while they themselves stood for the socialist ideals.

The head of department Voigt tried to keep up the appearance of infallibility in front of the workers. He concealed his lack of skills and his incompetence with authoritarian manners. However, the workers watched him closely and tried to find fault with him. One of the workers, Fröhlich, who manufactured the vibrating tables and who had come into conflict with the head of the workshop particularly often, closely observed his superiors Voigt and Dr Schöpf. He worked hard to reveal the failings in their ideological resolve. He thought about how their body language was associated with the fact that they made themselves out to be representatives of the truth without necessarily telling the truth. During speeches and addresses, he would count how often within a half-hour period they would say words like 'God' or 'believe', although as communists they had to be atheists.

Full of malicious glee at the head of department's expense, Fröhlich told me the story of how he had caught Voigt by surprise as he was manufacturing something in the factory for his own personal use.

> We were coming from the shop floor, going out the door. Voigt appeared from the corridor carrying a something that had just been chromium-plated. It was obviously something he had done for himself. Then he noticed I had seen it. I went up the stairs as if I was going to the toilet and he headed out the door. I thought to myself: 'You had better have a look what he's going to do now.' He went behind the glass door and stood still, checking that I had gone. Then once he thought he was in the clear, he dashed off in the other direction like a bat out of hell. Barely a minute later he was marching at full speed towards the train station, still carrying the part under his arm. Conceited to the extreme! … Well, making complete fools out of everyone else: 'I am totally clean!', he would always say, 'I dare you to find anything on me!' He was terribly frightened of anyone finding some flaw in him. (Fröhlich, skilled worker, Stanex, 16 June 1991)

The ideological control which Voigt tried to have over the workers achieved the opposite effect because the workers then judged him according to his own ideological standards. Anecdotes like the one I heard from Fröhlich were plentiful and were told again and again for the amusement of other colleagues. Such a criticism resulting from the ruling discourse itself is, according to Scott, 'the ideological equivalent of being chased up into the sky by one's own missile' (Scott 1990: 103). The type of criticism that was possible by applying the official discourse succeeded in unveiling the form of domination in its entirety.

The Freedom to Think Differently

The authority the party could have over its members was never absolute and was refracted through life's practical experiences. In this way, the view of the party held by individual comrades changed during the course of the political and economic development of the GDR: through disruptions in their personal life and events occurring within their social milieu. In everyday life, the SED members were not a monolithic block which the non-members had to face. However, they had this effect whenever they acted together according to party discipline. This was clearly demonstrated by a tragic conflict at the beginning of the 1980s that shook the department of design engineering of Stanex down to its very foundations.

In the early 1980s, an experienced engineer, Weber, who was not a party member, ran the Stanex design engineering department. Weber was the complete opposite of Schöpf, both in terms of politics and expertise. Colleagues in both engineering and manufacturing appreciated him. Nobody contested his expertise and this made him untouchable. In contrast, the people I talked to at Stanex described Dr Schöpf as someone who could not bear to be contradicted and who firmly stood by a decision, even after it had already been proven wrong. When Weber started confronting Dr Schöpf on a moral-political level, he posed a genuine threat to Schöpf's authority. Weber openly opposed Dr Schöpf's policy of abusing the monopoly the enterprise had for short-term enterprise interests. Most of all, he disagreed with how Dr Schöpf dictated contract conditions to his customers, shortening the guarantee period to half a year and delivering to them, instead of spare parts, design blueprints, which they were to use for producing the parts themselves. As Weber openly presented Dr Schöpf's actions as improper, he questioned the enterprise director's claim to absoluteness. Weber argued more in terms of national economy than Dr Schöpf himself, who deemed himself to be the idealistic guardian of socialist philosophy. Weber told Dr Schöpf how he believed it to be a waste of economic and intellectual potential not to assume responsibility for the service of the highly complex machinery and to disregard the likely consequence that a year later they might all be useless due to a lack of spare parts and inadequate maintenance.

Tensions often emerged between the director of the enterprise, Dr Schöpf, who referred to himself as a Stalinist, and the head of department, Weber, right up until 1982 when Schöpf succeeded in securing the two young comrades Kater and Fischer as design engineers. Both of them had studied together in Czechoslovakia and when they started at Stanex they were still full of political fervour which they had developed in their small party group in Brno. Immediately they tried to start up party activities and to agitate colleagues, but this was stopped by the non-members Weber and Scheuch. The conflict culminated in legal proceedings taken by the comrades against Weber and Scheuch, accusing them of being ringleaders. Scheuch told me how it all happened.

> When they first arrived, Fritz Fischer and Veit Kater were young comrades. We already had Niederegger and Karst with us, who were our comrades at the time. Well, these two young lads, they had the perfect upbringing, as we always said. They came well-prepared from home, they had studied abroad and had thought how wonderful everything is. So they came along to us and tried then of course to rouse everyone: party activities and all that. We opposed it right from the start. I wasn't head of the department back then, I was a design engineer. Weber was in charge. He couldn't put up with hearing all the hubbub either and we continued to oppose it a bit, and then legal proceedings were instigated by the four comrades against Mr Weber and myself. I don't know what they called it: 'Ringleader' I think. They claimed that they couldnt promote the ideas of the party because of us. (Scheuch, 24 June 1991)

Dr Schöpf used the legal proceedings as a pretext for forcing the unwieldy Weber to resign from his position as head of department. Scheuch, on the other hand, was to be transferred into 'an established collective' in the same enterprise. Neither Weber nor Scheuch could be dismissed officially on the grounds of ideological differences, but they could be relocated to positions that they were unsuited for and where they were dragged away from their friends, thereby isolating them both politically and socially. Dr Schöpf also strove to spoil Weber's future career by writing up a particularly bad reference for him. Weber faced the consequences and vanished, presumably fleeing to the West across the Baltic Sea.

> Mr Weber is no longer around. He has been a castaway since 1983. We all reckon he tried to escape, probably across the Baltic … and he was never seen again. … Well, he would have given us a call if he had made it. In the last thirty years lots of people tried to flee, but he mustn't have made it. (Grabher, process planning, Stanex, 7 May 1991)

The wall had been built because of people like Weber. His technical know-how was valuable for the system, but his stubbornness was undesired and this resolve was to be broken. Weber had not only refused to join the party, he further dared to judge the practice of those in power by their own idealistic discourses and then to expose them. Weber's disappearance was the talk of the enterprise for months and it meant Scheuch could be left in peace. His disappearance turned into a key drama for the whole division of production machinery in the enterprise and constantly turned into a bone of contention for political clashes within the enterprise. In 1984 Veit Kater became head of the design engineering group, but his relationship with Dr Schöpf continually worsened. It was possible that he also realized – what his colleagues were already discussing – that he had been exploited by Dr Schöpf for his own power struggles and had been used to solve Schöpf's inner contradictions. Kater began to question the conception that political work was possible and necessary.

> Well, nobody can be convinced in the practical sense. You can't really change an adult person anymore. Actually, not change them anymore, that's not putting it right. You cannot educate an adult or change his basic attitudes or do this through some sort of abstract discussion – I can't really imagine that this actually works at all. (Kater, design engineer, Stanex 25 April 1991)

Following a heated debate on New Year's Eve 1983, which nearly turned into a fist fight, Scheuch and Kater talked things over and this laid the foundation for a bond of trust that continued to last in the years that followed.

The design engineers' relationship with their enterprise was not a purely functional one. In the enterprise they attempted to realize their conception of meaningful coexistence and political ideals. The design engineers Scheuch and Weber tried to create in a society imbued with ideology an oasis where the only things that counted were skilled ability and creativity. The comrades Kater and Fischer tried hard to impart their recently acquired convictions to their colleagues and form a model social collective. The director, Schöpf, who had to fight for years in order to be able to produce machinery developed by his engineers, wanted above all to keep the sector alive and to be respected without question in his position as a socialist director. The clash of these diverse interests ended in tragedy because the comrades refused to reach a compromise with the non-party members. Yet, the loyalty of the party comrades was not rewarded with trust from the party. Instead they had to put up with suspicion from the party and, to make matters worse, they were not given the chance to justify or explain themselves.

Perestroika in the Enterprise

The key drama surrounding Weber's disappearance contributed to breaking down the rigid barriers between comrades and non-members. Gorbachev's visit in 1986 incited an even greater number of heated discussions in the enterprise and led to a division amongst the comrades between Gorbachevists and Orthodoxes.

> In this entire discussion since '85, ever since Gorbachev, the official line in the GDR has strictly been: whatever they are doing over there, it doesn't concern us – even as far as this famous line from Hager: 'If your neighbour is putting up new wallpaper, you don't necessarily feel the need that you immediately have to paper your own flat.' That sparked everything off. One side said: 'No. Whatever is going on there, it doesn't work here. It's no good, it's counter-revolution. God forbid that we let something similar happen over here.' And the other said: 'That's our only chance to get out of this mess, once and for all.' (Kater, design engineer, Stanex, 25 April 1991)

In the engineering department both of the young comrades Kater and Fischer now faced comrades Niederegger and Karst, whom Kater labelled orthodox or devout communists: 'The other two were people, who thought: we have to do exactly what we've been told to do. We can't start discussing it now. Consciously or subconsciously they turned a blind eye to change because they didn't want anything to be different' (Kater 25 April 1991). The debates with the non-members too became increasingly intense. As one worker from production perceived it, 'if they only had discussions up there, I don't know. It was war up there, a war of words.'

During the course of the 1980s the political discussions also became more and more open. The hidden critical discourse spread out into ever-growing ripples and the *hidden transcripts* gathered a growing audience.

> The enterprise militia was discussed, well, the job of the enterprise militia, the East/West border, the Wall, why … – and because you are saying 1988 – well, that was the time after we began to be represented in Bonn and Berlin. At that time the borders started to relax. Rising numbers of people were getting across and coming back. In any case, this provoked a lot of discussion of course. As the problem presented itself more openly and took shape more clearly, it was of course more intensely scrutinized and discussed. (Schuster, skilled worker, Stanex, 22 April 1991)

Debates even took place in the party group after 1985. However, these discussions had limits which were instinctively respected. Neither the 'state supporting ideology' nor the social and political system were allowed to be fundamentally questioned. Comrade Schuster told me:

> I reckon in the last few years there were more discussions inside the party than what actually surfaced publicly … there was always the risk, if I may put it like that, if anyone voices their opinion too loudly, perhaps they wouldn't be permitted to go abroad. So you had to bear this in mind. It was like that almost every day. So I definitely couldn't just stand up and say: 'Erich [Honecker; J.B.] is doing just the opposite – maybe not even the opposite – is doing it all completely different from what Gorbachev is doing. Because my view is that Gorbachev is doing the right thing, and well, Erich is making a pig's ear of it all.' I would never have been able to come forward and say that. (Schuster, skilled worker, Stanex, 22 April 1991)

The party members in favour of reform did not mix with nonmembers and avoided contact with opposition groups. This meant they were trapped within party discipline and were not able to act within the party either. While passionate discussions were taking place in the enterprise, to the outside world everything seemed to go along its socialist course as normal. The reformers thought a rapprochement that was too close to the bourgeois-western structures would endanger the GDR (Meuschel 1992: 309).

There was a restrained plea for perestroika (openness and transparency) in the enterprise. People did not wholeheartedly accept the SED party leadership's denial of the sweeping changes that were transforming the Soviet Union. In an incomplete brigade diary of the engineering department from 1988 and 1989 I found a letter dated from 28 November 1988, which the department, under the brigade name 'Collective for the friendship between peoples', had sent to the Society for German – Soviet Friendship. In the letter the collective ordered the Society for German – Soviet Friendship to make every effort to have the magazine Sputnik put back in the postal system. The engineers made it clear:

> Preventing any further publication of *Sputnik* in our country does not provide any solution, but is what we understand to be a motion of no confidence into our capacity for independent thought.
> We expect you to respond to this letter: we would also regard as a legitimate response the publication of the official position of the central committee of the DSF [*Gesellschaft für deutsch-sowjetische Freundschaft*, J.B.] in the press because on no account do we only read *Sputnik*.

This letter was the engineers' initial step towards addressing the general public, albeit only a small section of the public, and an effort to grant an open forum to the debates they had inside their department. All the same, they never received a written response to the letter indicating the DSF's position. Instead, the Society for German–Soviet Friendship tried to resolve the awkward matter by an informal chat, which would not leave behind any written trace. They answered:

> We have received your petition. A functionary of our friendship society will contact you for consultation.

The engineers' stance clearly demonstrates an eagerness for conflict and goes beyond the behavioural patterns of obedience and 'the delegation of political action to the "authorized people"', who Hanke (quoted in Meuschel 1992: 310) regarded as typical for GDR society in the 1980s. Nonetheless, this initiative gradually faded and, probably like many others in this period, was watered down by the bureaucratic mechanism which had standard answers prepared for subversive questions.

Not until the course of 1989 did GDR society enter a phase of open confrontation, which compelled the critical party comrades to take up a stance. For the engineer and enterprise militia member Kater, this moment had arrived when the members of the enterprise militia were being prepared several times throughout 1989 in order to be deployed against unarmed masses of civilians:

[The deployment of the enterprise militia was actually] a kind of strategic military game of chess. If the army had a front, and if the military opponent had deployed agents on our territory, then the enterprise militia was meant to take on the role of protecting our buildings. There were six to eight training sessions per year, and every two years there was the session called something like 'barriers and fences on streets and public places'. We were trained on how the enterprise militia was supposed to face a mass of people. In '89 we suddenly found ourselves doing this session three times in the same training year. Lots of us were hence suddenly asking: 'So what are we actually here for? Are we here to defend against an attack from outside or are we in fact here to stand against our own people?' Big discussions about this ensued and that was one of the reasons why I said: 'I'm not going along with this anymore, otherwise I'll suddenly be confronted with a situation where I will be facing my own people with a gun in my hand.' At that point I said: 'No, I'm not doing that.' (Kater, design engineering, Stanex, 25 April 1991)

His concerns were not unfounded for on 9 October 1989 in Leipzig, enterprise militia, albeit unarmed, were deployed against demonstrators. The workers of the Leipzig Kirow works found themselves on both sides of the barricade (Hofmann 1995: 183-84). Kater left the SED shortly before the Wall fell when he ran the risk of having to defend the ideas and interests of the party against the people through armed force. Even waiting until this late stage before quitting the party did not merit him an honourable departure. Instead, the party retaliated by excluding him.

In spite of their intensity, the debates in the engineering department and party organization of the enterprise remained 'hidden discourses', which did not surface outside the enterprise. Even the critical members thought it was only by means of the party that they 'could achieve something in society' and that they could only be critical inside the party. By not letting criticism penetrate beyond the inner circle, they maintained the party's claim to absolute truth, in which they also had a part as members. Decisions like leaving the party were not reached collectively, but were left to the individuals themselves. The reformers or Gorbachevists did not split from the mother party or appealed to the general public with an alternative social programme. The hidden discussions within the party therefore had no consequences for the radical transformations in the GDR, which finally broke free when large sections of society began to take care of their own interests themselves (Meuschel 1992: 309).

After the Wall fell, the West German reformers of the planned economy overlooked both the party's role in the enterprise and the stranglehold of politics on the GDR economy, and they therefore simply disapproved of this as an ideological burden. The directors of the people-owned enterprises, who were involved in negotiations, were either valued as pragmatists or were derided for lacking the necessary skills.

In the stories the employees told about the power relations in the people-owned enterprises, there always lingered a need to bring the political paternalism to account, which they had tolerated in their job. In the eyes of most employees, the political role played by the directors in the planned economy disqualified them for an economic role in market economy. When in autumn 1990 the old directors were confirmed by the Treuhand or the Western buyers as the managers in all three enterprises, this in fact took place against the explicit request of most employees. It also affected my enquiries because employees now stressed that they 'could not tell me anything' or 'could not say everything' about the political role of their superiors in the past, since these were still in the same places and positions. They thus had to be careful not to endanger their jobs by disclosing information too openly.

Notes

1. 'Entschiedenheit an die Stelle von Entscheidung setzen'
2. 'Zu verbessern gilt es die Qualität des Mittagessens.'
3. 'Die Qualität des Mittagessens muß jedoch noch gleich bleibend gut gewährleistet werden.'

Part II

The Wende

In the final years of GDR rule, as the debates showed in the party groups at Stanex, the developments in the neighbouring countries were accurately perceived. They were not discussed openly but provided material for arguments privately and at work. In particular, the Gorbachev reforms in the Soviet Union were fiercely debated. In all three enterprises the Society for German–Soviet Friendship received a new lease of life, while members of the SED left the party in droves during the exchange of the membership books in 1989. Veit Kater was among those leaving the party in September 1989. Open political resistance remained absent until autumn 1989, but a noticeable retreat from organised 'social' life was evident.

The 'possibility to travel' (*Reisen-können*) was an important subject of conversations in the enterprise. The right to leave a country is, as John Locke expressed it, a particularly significant right for only if this right can be exercised can conformity with a regime or system become a matter of choice (Lukes 1990: 30). For some of the people I talked to, a failed application for travel to the West was the catalyst for renouncing all official functions that could be taken to be political support for the SED regime. Since he was not granted permission to travel to West Berlin for the birthday celebration of a relative, the engineer Born renounced his function as shop steward, quit the Society for German–Soviet Friendship, left the Chamber of Technology (*Kammer für Technologie*) and refused to volunteer any longer on the board of the Housing Community. In the written explanation he submitted for his resignation, he wrote that as he was not trusted to return from a journey to the West, he could no longer hold such positions of responsibility. Other colleagues developed an increasingly critical discourse towards the state. As Fröhlich reported, during the shop stewards' general meeting shortly before the Wende, members stood up and protested that in this country they were left 'to stew in their own juice' and could not achieve anything, because they were kept ignorant of 'what was going on in the world' (Fröhlich, skilled worker, Stanex, 16 June 1991).

For forty years the social reality had been interpreted, defended or criticized from the perspective of the isolation of GDR society. The normative principles, which the opposing factions in the three enterprises used as the basis for their arguments and ideological clashes within the enterprises in autumn 1989, were a product of the planned economy and of the socialist order of society. The principles were diverse, highly moralistic and above all scarcely oriented towards concrete strategies for action. As long as the controversies stayed private, the possibilities for real changes remained small.

Not until summer 1989, when people saw the possibility of emigrating, did they gain the confidence to consider other options, such as reacting to an intolerable situation not by moving away, but by attempting to transform it through dissent (Hirschman 1992: 344). The motto that was frequently proclaimed in autumn 1989: 'We are staying here!' became an ambivalent but effective threat to the GDR authorities because it proved that the tacit acceptance of the social conditions was now over.

For a short period in autumn 1989, hopes were pinned on the trade union with Harry Tisch at its head because it seemed to form the only alternative force to counter the Socialist Unity Party. On 20 October 1989, the shop stewards and trade union leaders for the section assembly automation at Stanex wrote a letter to Harry Tisch to voice their support for a public letter, which the trade union members of Bergmann-Borsig had aimed on 29 September 1989 at the trade union directors. The trade union members demanded a public enquiry into the floods of GDR citizens pouring into the West and a discussion of the real reasons behind their emigration. While avoiding questioning socialism as an ideal of society, they insisted: 'We have to offer people new perspectives which enable the further development of what has already been achieved on the basis of real individual exertion of influence.' Hardly two weeks later, the man they had 'pinned their hopes on' had to resign from his post of FDGB chairman and one month later was accused of having grossly abused his position.

When the wall fell, the old regime not only collapsed, but discourses that were critical of the regime also became rapidly disoriented. Most of the people I talked to in the enterprises stressed that in 1989 they did not want to do away with 'socialism' as a social ideal and moral order, but that they defended themselves against the political and social patronizing that accompanied it, as well as against the irrationalities of a centrally planned economy. The plea for individual liberty, freedom of expression and democracy was overtaken by rapid political and economic developments, which were not brought along by any organized revolutionary opposition, but were managed by the old party elite. There were 'no counter-elite, no theory, no organization, no movement, no design or project according to whose visions, instructions and perceptions the breakdown evolved' (Elster et al. 1998: 11). Elster and his colleagues label the dissolution process of the old system strangely 'subjectless' and the process of setting up the new system 'vegetative', hence extremely passive (1998: 15), because they could not discern any social agents, and above all any social elite, who would have driven the process forward towards specific goals and intentions.

While in most postsocialist societies the power relations within institutions remained unclear after the collapse, in East Germany institutions of the Federal German State filled the empty space. Through Federal German law and institutions, the members of East German society received a new legal framework, which on the one hand placed limits on arbitrariness, but on the other hand was not founded on their own initiative. The process whereby institutions were transferred from West to East is sometimes erroneously termed 'colonization' (Stark and Bruszt 1998: 175), which, despite certain limits to its democratic character, does not do justice to German unification and serves to play down the violent subjugation and exploitation of large parts of the world by colonial powers. West German society, which was no longer seen by GDR citizens as merely an ideal model to follow or as a frightening polar opposite, took the place of a social alternative and grew into the dominant and domineering model.

East Germany saw its future being shaped by West German politicians, who had infinite trust in the all-encompassing market principle and confidence in the powerful Federal German State, but who were, however, sceptical towards East German society (Stark and Bruszt 1998:102). Also the privatization trust *Treuhand*, which was created by the Modrow government in spring 1990, at first left privatization to the market and sold off East German enterprises without restructuring them beforehand. Yet, in 1990 it was not investment capital that flowed from West German firms over to East Germany, but goods, which quickly stifled East German products and resulted in a 'market shock' (Stark and Bruszt 1998: 138). The East German enterprises lacked managerial experience, access to markets, new technology and moreover could not compete due to rising wages. As a result of this, 4.5 million jobs were lost between 1989 and 1992. From the first six months of 1990 to the first six months of 1991, industrial production sank to half its original level, and a year later it continued to fall to a third of the level in 1989. The Treuhand did not begin to change its approach until 1991 and only then did it assume responsibility for everything the 'invisible hand of the market' was unable to achieve: safeguarding jobs, preventing deindustrialization, securing the ability of private firms to survive and so on. The active economic restructuring of East Germany and the transfer payments from social insurance turned into astronomical costs for the citizens of the Federal Republic of Germany. Stark and Bruszt (1998: 140) talk of payments being twelve times more than the Marshall Plan.

Many citizens in West Germany felt the economic and political consequences, in particular the costs of the state debt and tax increases, to be a considerable burden and an involuntary tribute towards unification. Furthermore, the relations between East and West Germans, and especially East and West Berliners, were moulded by intimate bonds, which gave the political and economic relationships a personal, emotional dimension.

In this transformation process, the employees in the three enterprises pursued individual strategies and various moral and economic goals. Every single one fell back on knowledge acquired in the planned economy, formal and informal relationships with colleagues, superiors and customers, recollections of past wrongs and loyalties. All the same, people did not remain the same while the institutional order collapsed around them, but they changed, experimented or opposed and contributed in some measure to the society that arose during this period. Nonetheless, the outcome did not necessarily correspond to the original intentions. The Wende in the three enterprises is also a story of objectives that were never fulfilled. Why that was so, I will show in the two enterprises Taghell and Stanex, which tried to survive in the market economy without being bought up by Western firms. What I find particularly interesting here is the view of the market economy, which the employees developed, their conceptions of right and wrong and the instrumentalization of money and ownership in the new power relationships.

Chapter 5

Privatization – Domination and Possession

Instructions once again

He loves them, the people, the omnivores
He cooks and they gobble it up
He makes it clear, none of them could beat him
And often comes through the house.

The people are weak and it hurts them to think
They are happy that he's doing it for them
The eyes – they shine so brightly – oh!
And deep in your head it is night.

They plug away, they toil, they fight for him
He shouts and they come running
If men don't withdraw from the Man
Their field of life lies bleak and fallow

Look for the enlightenment switch in you
Give your inner windings some light
Never eat convenience food with beer
If not you'll remain a piteous gnome

Horst Froberg (alias Fröhlich, skilled worker, Stanex, 1992)

To whom should the enterprise belong? Or rather – who will take responsibility for its fate? These questions are the background of many fights that took place between 1989 and 1991 in two of the enterprises examined. Plant managers and employees vied for control of the enterprise. While the managers endeavoured to acquire exclusive ownership rights, the employees did not raise individual exclusive claims to possession – rather they sought to be involved in the enterprise's reorganization and the economic and personnel-related decision-making processes. at the beginning of 1990 it seemed to go without saying that, following the fall of the Wall, they had a pivotal role to play in the changes taking place in 'their' enterprise.

In the third part of this book, I will contrast the internal conflicts I was able to observe here over a two-year period with the privatization of the third enterprise by a multinational concern – 'from the outside', so to speak. The integration of the third enterprise into a concern that operated on a worldwide scale worked somewhat differently and was heavily shaped by global corporate management strategies and the influence of a strong workers' committee.

The Treuhandanstalt's (trust fund's) concept of privatization allowed neither for the workers to share in the possession of the enterprise, nor for their participation in the privatization process. Its aim was the allocation of all property in the GDR to private legal entities. This corresponded with the prevalent notion that this would lay the foundations for a healthy society founded on civil and political freedom and economic performance. Private ownership would be superior to collective ownership, because private owners would be more motivated to tend to their properties than if they had a part in a collective property (Hann 1998: 14, 17).

The managers of Taghell and Stanex both wanted to become enterprise owners. Through ownership they would obtain exclusive control of decisions, equipment and access to external agencies such as the Treuhand. The liberal interpretation of the word 'possession' gives the owner the right to exclude others from using or consuming a material or immaterial object and, with regard to the enterprise, the right to decide who belongs to it and who doesn't. In the process of German unification, the right to access the enterprise and to control the access of others posed one of the pivotal political problems of democratic control (Hann 1998: 46).

A strong tradition within social anthropology has come to define possession as 'the relationship between people with regard to things' (Verdery 1998: 161, Hann 1998: 4) rather than regarding a possession as a thing, or as the relationship of a person to a thing. This viewpoint allows the privatization of an enterprise to be understood as a process, in which power relationships are re-negotiated. Through this process, possession becomes a 'bundle of power' (Verdery 1998: 161).

In the case of both enterprises, the managers acquired economic entities of undetermined monetary value. It is possible that in 1991, they already belonged to those enterprises with 'negative value', which the Treuhand tried to privatise and preserve (Stark and Bruszt 1998: 139). More important than current market value was the social position of 'the owner', the power over others it conveyed, and the legal right to support and subsidies, which the Treuhand and the Federal Government accorded new owners. Private, exclusive ownership cemented an immense power imbalance, which was already in place in the formal hierarchical structure of the planned economy and which gained in potency following privatization.

In order to understand the consequences of such exclusive allocation of power for the people in the enterprise, it must be examined how legal regulations are inscribed in everyday enterprise routine, in values- and moral perceptions, and in power relationships. Personal, political, legal and economic factors are inextricably entangled in the privatization process, which reduces the simple premises of liberal market ideology to absurdity.

Covert Criticism becomes Loud

At the first workforce meeting following the fall of the Wall, directors had to face questions about their capability to lead the enterprises. At the end of January 1990 at Taghell, on the initiative of the workers and foremen of the polishing workshop, a discussion took place between the enterprise director, the party secretary, the representative of the *Betriebsgewerkschaftsleitung* (the enterprise council of the trade union) and two directors who were part of the executive body. The minutes of this meeting, (26 January1990), as taken by Kaiser, the foreman, detail the questioning of the management about technical and financial aspects of the enterprise. The workers wanted to know whether their enterprise was profitable, why so many lamps had already been stockpiled before the fall of the Wall, how the production plan for 1990 should be achieved and whether the enterprise could get out of the Kombinat (the socialist conglomerate). Two diametrically opposed points of view emerged at the meeting. One of these was that of the enterprise director, Fechner, who vouched that the enterprise would function at a profit and achieve the targets set out in the plan. The other was that of the old foreman Kaiser who was still able to recall his experiences of the time before collectivization, and who declared that the plan was fulfilled only once the merchandise was sold. While the director continued to argue in terms of a positive production balance and plans that could be fulfilled, the foreman demanded a critical appraisal of the chance the product would have in the marketplace. He still expressed himself in terms of fulfilling the plan, but the content of his criticism was directed squarely at coping with the new market era.

Stanex employees reacted in a similar way. In the spring of 1990, a small group of skilled workers and design engineers from assembly automation tried to expel the area director, Dr Schöpf – with the help of the head of the enterprise, Mayer. The skilled worker Walter Schuster described their attempts:

> There was a clique of conspirators … We organized ourselves and said: 'That's it. We'll get rid of the Dr!' [the director of the section; B.M.] … Over at the meeting of the entire staff we talked about it with Mayer [the head of the entire enterprise; B.M.]. Mayer said: 'What you want is all fantasy, capitalist nonsense.' (Schuster, skilled worker, Stanex 22 April 1991).

Enterprise head Mayer, who had had sympathies from the workforce and had been on friendly terms with some of the design engineers in the automation section before joining the party and making a career for himself, stood by area director Schöpf and did not agree to his removal. He attempted to hold the large, seemingly futureless, enterprise together and, as one worker commented, obtain state support for it.

> November/December, January/February! The first passionate discussions took place. The suggestions had already been made. But the leadership style in our enterprise was still a socialist one. Known as the 'Luftish line' [a reference to the political views of Christa Luft, the Modrow Government's Minister for Economic Affairs; B.M.], the enterprise was to continue with the help of state support. We'd already told him that something like this wouldn't work, you don't get this kind of thing in a capitalist society, in the social market economy, as they falsely call it! You can only make social security provision for employees in a market economy when that economy is rich enough. A market economy means survival of the fittest – the best, the strongest wins the contest. (Schuster, skilled worker, Stanex 22 April 1991)

Stanex's workforce also seemed better aware of the requirements of a competition-based market economy than their old leadership. However, their vision of fair competition did not really coincide with current practices of market economy either (see Chapter 6). The politics of the directors was shaped by a worldview from the planned economy and remained reactive. They tended to wait for instructions from above, from the last GDR government headed by Modrow, from the Ministry of Economics, and later from the Treuhand.

The lines of attack of the workforces of Taghell and Stanex reflect the management styles that were prevalent before the fall of the Wall. The leadership of Taghell's director, who had not exerted any ideological control over the enterprise and had held off political pressure from the workforce – but who had in the opinion of the employees enriched himself personally – was called into question on a practical level. Conversely, the workforce wanted to remove Stanex's director on account of his 'Stalinist' leadership style.

The employees explained to me, with hindsight that they had thought it obvious that the end of the planned economy would entail the fall of the old directors. They had also believed that a much more important role than before would fall to workforce representatives when it came to transforming the enterprise. The directors themselves were so unsettled by this development that they preferred to give any argument with the workforce a wide berth. In the spring of 1990, when the most important decisions were due to be taken, Stanex's director obtained a sick note for a number of weeks. Taghell's director refused to set foot in the workshop.

At the staff meetings that took place between November 1989 and April 1990 radical demands were not made, nor were the few resolutions that were agreed on actually put into practice. However, these meetings remained firmly engraved in the memories of those involved for years. The employees, speaking out against their directors, were for the first time able to raise questions that had only ever been whispered during the time of the planned economy. The fact that for the first time they put into words what they had thought during the time of the planned economy filled them with a sense of pride – and deeply unsettled the directors. It filled them with a sense of achievement and satisfaction that they no longer had to feel subservient to their despised bosses (Scott 1990: 208–9). However, even the most active workers' representatives were not prepared to take on the role of the directors, since they also had no idea of how the enterprise should develop in the future. Three years later, the skilled worker Schuster gave this résumé of what had happened:

In the end the younger members of the workforce were even more paralyzed than the older ones. It totally stunned me when they said: 'Not us, for heavens sake!' No group was formed that could have come up with an alternative suggestion. Those who wanted something new, did not want to go against individuals personally or set their actions up politically. They just wanted to get rid of the directors, because according to our opinion, to the opinion of a few people, this was the condition for improving work here. As no alternative arose, individuals were not challenged and no action was taken against them. (Schuster, skilled worker, Stanex, 20 April 1993)

The directors' fears of losing power therefore bore no resemblance to the workforce's actual willingness to spur on their dismissal.

From 'People-owned' to Privately Owned Property

The privatization institution, Treuhandanstalt, set up by the Modrow government, became the symbol of Western dominance of the East, the GDR's economic downfall, the transfer of ownership to West German citizens and of East German 'incapacitation' by 'arrogant' West German bureaucrats. During the critical privatization phase, however, Stanex and Taghell employees regarded the Treuhandanstalt as the last regulating authority they could appeal to. 'We should go to the Treuhandanstalt', was a phrase heard time and again, particularly when cases of enterprise mismanagement became evident and when the employees had grounds for suspecting that their manager was enriching himself at the enterprise's expense.

In 1990 the Treuhand exerted its greatest influence over the power relationships within enterprises, when it endorsed the employment of the old directors of people-owned enterprises as managers of the newly founded GmbHs (*Gesellschaft mit beschränkter Haftung – Plc*). This gave them the institutional sanction of their position of power and formally transferred the hierarchies of socialist enterprises to the newly founded companies of limited liability. The Treuhand acted as the kingmaker during the transitional period leading to privatization, when institutionalized forms of power were the most useful resource in the struggle for power and prestige between individuals and groups (Lenski 1986: 249). The managers were subsequently able to use their position in the contests for jobs, shares, enterprise assets and wages.

The dream of quick money

Taghell's privatization both cemented and strengthened power structures that had already existed before the Wende. Before the Wende Fechner, Taghell's director, had based his enterprise's success on dealing brass lamps against spare parts and materials, and on relationships of favouritism, on 'mutual gifts of pleasure' as it was called. His departmental directors had joined the enterprise because there was 'something to be had'. As Kaiser and Saller told me, they had been dealing in brass scrap and lorries full of lamps had left the warehouse late at night for some unknown destination.

The enterprise politics that Fechner adopted following the Wende, when he had understood the weak points of the privatization politics supported by the West German state, remained in line with his strategies in the planned economy. Only the control mechanisms, which the social market economy used during the transition phase, were essentially less effective than the control exercised through the plan and the party during the period of the planned economy. The Betriebsrat (workers' council) of Taghell had fewer opportunities for control than the conflict commission had had before the Wende.

In the years between the currency reform of 1 July 1990 and the day when Taghell went bankrupt on 1 September 1992, the enterprise was home to covert and overt power struggles, intrigue, and shady deals. As the widow of the former owner still held a 2.4 percent share of the firm, the Treuhand's reprivatization department dealt with Taghell. When it came to calculating reprivatization claims, the unification treaty guidelines provided different formulae from those governing the sale of public property to private owners. The former owner became a GmbH partner together with the Treuhand and was able to request reprivatization – which meant he/she got his/her property back and could claim compensation. The partnership agreement, with the Treuhand as one of the shareholders, stated that all decisions were to be taken unanimously. Also the managing director could only be revoked if all partners agreed.

In the course of 1990, the managing director came to understand the advantages he could extract from the reprivatization process and, without informing any member of the enterprise, he bought from the widow of the former owner her 2.4 percent share of the enterprise for a sum of DM 5,100. He hence became a shareholder and was now no longer revocable as a managing director without his consent. At the same time, a group of employees had come together in order to vote out the managing director. The ballot did not take place as planned because at the very last minute, the conspirators discovered that Fechner had become an owner. Mrs Martens, the shop steward, explained:

> Because someone had their ear to the ground ... He warned us and said: 'Don't do it. You really can't do it any more. Mrs Sanders has transferred everything over to Fechner. (Martens, Taghell sales department, 30 July 1991)

On 27 January 1992 Fechner also succeeded in acquiring Mrs Sander's reprivatization claim, which permitted him to apply to the Treuhand for the reassignment of the state-owned share of the firm, and to demand compensation for the deterioration in the enterprise's profitability since its nationalization in 1972. Fechner demanded a sum of 11.6 million Deutschmarks. The Treuhand made a compromise offer of 1.7 million Deutschmarks, which he chose not to accept.

Fechner fought a constant battle against the workers' council, which, in the autumn of 1990, with Mrs Martens as president, resisted the first lay-offs that Fechner had planned. When the trade union began to arrange training sessions for workers' council members and to make them aware of their rights, Taghell's workers' council had already lost its best representative and had been made fully dependent on the manager. Before the November 1990 workers' council election Fechner organized a scam that resulted in Mrs Martens being unable to be nominated as a candidate. He appointed both her and her colleague, Mrs Gertz, the manager of material management, as members of the managing body and then dismissed them again from this position two days after the election. Mrs Martens commented:

> At a meeting with the boss upstairs, we found out that we now belonged to the managing body. At that point I hadn't really grasped just exactly what was going on. Maybe I was also feeling a bit flattered by it all at the time – it might have been so ... – until I spoke to my husband about it at home later and he said: 'It's a trick, you know. Now you won't be able to run for election any more and you won't be able to vote yourself either.' And with that I was out. (Martens, Taghell, 30 July 1991)

The newly elected workers' council saw its role as being one of mediation between the plant manager and the personnel. Mrs Brandt, the new workers' council president, was also successor to the personnel resource manager, Kabel. In this double role, which had been common within socialist enterprises, she was present during the discussions of the managing director with members of the staff without knowing exactly which of her roles she was expected to perform.

> So then I don't know. Am I there as part of the workers' council or as a staff manager? I take the lists in with me, all the documents we work with. But then up pops the question: 'What is the workers' council's opinion on this?' Yes. And then I say: 'I can merely give you my own personal opinion. In order to take a decision the entire workers' council would need to be consulted.' (Brandt, personnel resource manager Taghell, 8 August 1991)

Mrs Brandt was pleased that in her role as personnel resource manager she was able to obtain snippets of information about planned dismissals or applications for subsidies on redundancy indemnifications. She learnt nothing about the negotiations between the Treuhand and the managing director, or about his privatization strategies. The members of Taghell's workers' council explained that they had steeled themselves to go to the Treuhand to inform themselves and to complain. They had however decided against it, as they suspected that members of Taghell loyal to their boss and with good personal connections to employees in the Treuhand would have informed him about their visit even before they had left the house.

> But look here – the *Treuhand* members are all old mates of his from the Economic Affairs Council (*Wirtschaftsrat des Bezirks*, WdB). They're from the Economic Affairs Council, they're from the Planning Commission, and they were his contacts in the department of external trade. The people he's done business with for years, are basically sitting there. He even once said himself in a meeting that he already knew everyone present, they were all his old friends from the WdB. (Martens, Taghell sales department, 30 July 1991)

In the eyes of the employees, the Treuhand was a reincarnation of the controlling bodies of the planned economy, where the enterprise managers and administrators who had been former party members mutually protected one another. The so-called 'little man' was left feeling as though he had no rights whatsoever, since he did not know what his rights were and since he had to worry about his job.

The managing director used the isolation of his workers' council to appropriate the firm's real estate property and to systematically destroy the enterprise. In spite of new orders from a major Swedish customer, from 1991 onwards the firm was faced with a catastrophic economic situation.

In July of 1992 Taghell's case was taken over by a new West German Treuhand employee, who attempted to obtain Fechner's 2.4 percent private share which was preventing the firm's sale. Only after being threatened with a charge of fraudulent bankruptcy was Fechner prepared to give up his share – on 28 August 1992. By way of compensation, he was entitled to guaranteed employment until the end of 1993 or to the payment of the equivalent corresponding salary, as well as to 15 percent of the revenue from the liquidation of stock and machinery.

In spite of the fact that Fechner had been dismissed on 28 August 1992 and a liquidator had taken over the affairs of the enterprise, the latter did not intervene until the end of September. Employees of the administration of Taghell left uninformed gained the impression that by 31 August something was bound to have happened, since a number of procedures needed to be completed by that date. In September of 1992, Fechner declared gloomily that he was now just an employee like everyone else. Countless letters were written and backdated to August. The letters were scattered among the different offices so that their respective typists were unable to make rhyme nor reason out of the whole procedure. During this period Fechner sold the cars to his former employee Bürger for just one Deutschmark and transferred to him the computer system that had cost 25,000 Deutschmarks. Employees also claimed that lorries full of lamps had left the yard for some unknown destination. Like a puzzle, the employees pieced together the elements of Fechner's enterprise policy and at the end of September finally plucked up the courage to send a delegation to the Treuhand. It was here that they discovered that the enterprise had already been liquidated on 3 September 1992.

Notified by the Treuhand, the liquidator appeared. He was now responsible for the damage Fechner had caused and had to take appropriate action. At the beginning of October he entered the enterprise accompanied by members of the public prosecutor's office and the police. On 6 October Treuhand members and the liquidation firm came into the enterprise with a team of five lawyers. In the presence of the lawyers, Fechner was accused of, amongst other things, taking 260.000 Deutschmarks of the enterprise's funds and selling the cars for one Deutschmark each. In his defence, Fechner referred to his discharge as a managing director signed by the Treuhand on the 28 August 1992 and pretended that all these events had already happened by August. While the liquidator spoke to his lawyers, Fechner drove the car that he had sold to Bürger for a Deutschmark from the yard. The Treuhand employees tried to explain to him that the discharge was only valid within the bounds of the commercial duty of care, which he as a manager should have observed. As he continued to deny any fault, the liquidator pulled a criminal and civil prosecution from his bag and banned Fechner from the premises. Taghell was dissolved as a firm on 31 December 1992.

The drama surrounding Taghell's privatization appears to be an extreme case, though in actual fact it is not unusual for the 1990–92 period, when accelerated privatization of the East German economy was taking place. Fechner, the plant manager, is a small fish in an ocean of both small and large economic scandals, which shook the Federal Republic after the Wende. Taghell's story is of no particular interest to me as a criminal case, rather than as an example of the mechanisms of power and powerlessness that come to the fore.

Because the employees were unable to evaluate the rules and laws that the new corporate and political institutions were abiding by, and understand why it was the functionaries of the planned economy who were conducting the transition into the market economy, they as a consequence distrusted the institutions that had survived unification – the Treuhand, and the trade unions. They hesitated when it actually came to demanding their rights of control and participation because the Treuhand had always appeared impersonal, dangerous, and even mysterious to them. The managing director, on the other hand, was able to rely on an efficient network of long-established connections that had continued into the market economy in order to circumvent obstructive rules and laws. He quickly identified the weaknesses within the privatization laws and used them to his advantage. In his excess, however, he ignored the fact that the more institutionalized the Treuhandanstalt became, the more it refined its corporate control mechanisms, which eventually weakened the networks that Fechner had relied on. The flagrant abuse of the appropriation strategy that he had constructed eventually caused its collapse.

The enclave in the world of market economy

During my first visit in June of 1990, the employees of Stanex's assembly automation section were already discussing the possibility of becoming self-employed and of separating their section from the main enterprise. They had not imagined that they could lay claim on the people-owned enterprise as their private property nor could they imagine any other legitimate owner either. They spoke about 'their' enterprise that they had helped to build, and about 'their' product.

> We thought our product was a good one too! We think it's absolutely brilliant – and we know we can carry on producing something similar! We'll stand by it! (Schuster, skilled worker, Stanex, 22 April 1991)

When the Treuhandanstalt announced in December 1990 the directive that the sections of the enterprise should be restructured to make them attractive for potential buyers, the debates among the employees intensified about whether to take possession of their section through an employee buy-out. However, the dissolution of the main enterprise also meant that the individual sections would have to take responsibility for the profitability of their production. The illusion of a protective wall separating them from the harsh reality of the market was now definitely a thing of the past and the workforce was painfully aware of the fact that sales orders were critically absent.

In January 1991, as the staff pondered about their opportunities for independence and becoming self-employed, Dr Schöpf, the section manager, took the initiative and distributed questionnaires to all of the employees. These were to be completed in detail, including employees' names. He wanted to find out whether they would be ready to become silent partners of an independent new firm that he would create out of the automation section of Stanex. The questions that he posed were directed towards finding out exactly how much money his employees would be willing to invest, whether in 1991 they would still want to continue working for the firm and whether they would accept him as their manager. Fifty percent of the employees were prepared to become silent partners, although the majority were unaware of the status this would have given them. The remainder regarded the questionnaires as a means of weeding out disobliging employees and complained about the fact that they had not been able to complete the questionnaires anonymously. Schöpf himself viewed the questionnaire as a kind of 'referendum' that should confirm him in his role as manager. As he said himself, he wanted to avoid 'being shot from behind'. In this way, the employees were unable 'to stab him in the back', nor to blacken his name by denouncing him as 'a red sock' (an active communist). He also wanted to safeguard himself from the workforce members who wanted to play an active part in the enterprise takeover – or, as he said: 'Those members of the workforce who have expressed an interest in democratic socialism in the past now aspire to people's rule.'

Born, an employee from project planning who was extremely critical of the questionnaire, explained his views to me:

> And then he wanted to know if we'd be prepared to invest money. At the meeting he'd already told us: 'But that doesn't mean you just leave your money like you do at the bank. If you think it's easy money, if you think you can leave the money to work for you, you'd be better off taking your money to the bank – they'd pay you more in interest'. The money that we were to invest here would just disappear at first. Even when things were going reasonably well we'd perhaps get 4 percent back.
>
> At the same time he warned us that if we thought we could denounce him as a member of a communist gang and abuse him in that way then he wouldn't put up with it. He told the guys squarely, he was a Communist and he would remain a communist – he would never become a part of the market economy. He'd also already warned us that: 'If you commit yourselves to me then you have to do it with your body and soul, completely ...' (Born, project planning Stanex, 4 February 1991)

The questionnaire's predominant concern was whether or not the employees wanted to become silent partners. Dr Schöpf wanted a declaration of belief and loyalty from his employees. They should place their trust in the product; they should be prepared to demonstrate their solidarity by investing money they might not see any return on. They should believe in him as their undisputed socialist leader and renounce payment in times of crisis. In order to become a limited partner in the kind of firm Dr Schöpf was striving for, one did not have to put up money for economic reasons, rather on the contrary: on the basis of idealistic conviction. Now that real existing socialism in the GDR had collapsed, Dr Schöpf, wanted to achieve 'Socialism in one enterprise'. He himself would personify the undisputed communist principle.

As Dr Schöpf could no longer expect that his employees would demonstrate their trust in the principles of real existing socialism, they now had to believe that their product was in great demand. This declaration of belief was contrary to actual fact, as since the currency reforms had taken place, the section had received very few orders. However, even Born, who was very critical of Dr Schöpf, thought it was self-evident that one would answer 'yes' to the question whether their product was in demand. His colleague from production, Schuster, later explained that Dr Schöpf had already in socialist times 'tied up a parcel': insinuating that the roundtable assembler as a special product, he himself as an exceptionally successful leader, and socialism as the only acceptable form of society belonged together.

For Born there was 'no question' that during difficult periods he would 'pull on one string' together with his colleagues. This appraisal was all the more remarkable as, at the same time, a number of employees, and Born himself, complained about the isolation and the egoism amongst colleagues. Dr Schöpf also wanted to know whether the employees would be prepared to put up with pay arrears in order to overcome difficult phases. He did not leave employees with the illusion that by investing in the firm they could earn money, rather he stressed that, at first, their money simply would not be at their disposal any more. This was completely in keeping with the fact that he portrayed himself as a staunch communist, whose political persuasion had not changed and who would not become part of the market economy, even if he were to act as an entrepreneur. For this reason it was merely consequential that he refused to make capital work and to pay out large profits on capital investments even at times when the business would be flourishing.

Dr Schöpf wanted his employees to commit themselves body and soul to him and to accept him as their leader. This would realize a situation that he had aspired to during the period of the planned economy. At that time, however, his power was not complete, as he constantly had to render accounts to superiors, both from the enterprise and the party. His staff was also aware of their position of strength as a qualified and sought-after workforce. By asking his employees to individually complete the questionnaire he succeeded in isolating them, since they could not know what each other's answers had been. This also corresponded to one of the tactics that he had employed during the time of the planned economy. At that time he had been successful in driving out charismatic figures from the collective – namely, the design engineer Weber from construction design, and the production technician Steffan.

Born spoke out against becoming a silent partner. He had planned to take early retirement a year and a half later. Above all, however, he was not prepared to invest money in a business led by Dr Schöpf. He was also reluctant to tell Dr Schöpf that he would go without wages and work during times of weak orders. For him, an enterprise run by Dr Schöpf was not the kind of business that he would have been prepared to make sacrifices for. It was also clear to him that as a silent partner, he would have absolutely no say and be completely at Dr Schöpf's mercy. His colleagues, in particular those who were even less familiar with the practices of the market economy than he was, were, according to his opinion, completely unaware of the consequences their answers could bring.

But to give something like this to a worker, a lathe operator, a transporter. They didn't have a clue what he was talking about! And then he wanted to explain what it meant to become a silent partner, the rights and duties and so on that you'd have. Absolutely no one understood – the people just sat there in disbelief, staring into the void, shrugging their shoulders. (Born, project planning Stanex, 4 February 1991)

In reality, around half of the employees had placed their trust in Dr Schöpf and offered him a sum of 130,000 Deutschmarks to invest. Although investments tended to be more forthcoming from those employees who feared for their jobs and who, due to their age, had few other opportunities in the job market, Schöpf did not want their investments to constitute a promise of job security. Moreover, he was keen to lead the enterprise alone, to avoid, in his own words, 'mass discussion about every little decision to be taken'.

In the first few months of 1991, the ambitions for self-employment were not realized and the section's order prospects barely improved either. In April the section made absolutely nothing, and in May it took in just 11,000 Deutschmarks. Using the 720,000 Deutschmarks that were due in July 1991 in payment for the sale of three roundtable assemblers as collateral, the combine had already taken out a loan. In May the Treuhand gave a Swiss real estate company the goahead to take over both sections of Stanex located in Berlin, under the condition that 241 employees, employed in four sections – one of these the assembly automation section – would be kept on. 1,337 were to be made redundant. Those remaining sections of the business outside Berlin were sold to other firms. In an agreement of interests between the workers' council and the managing body, it was established that sixty-two of the eighty-six employees from the assembly automation section would be reemployed.

In order to rid themselves of the responsibility for the 241 employees as quickly as possible, the real estate company offered the section managers the opportunity to hive off and to take over their section together with the employees. But some of the more critical employees also had an interest in taking over the section. Kater and Grabher contacted a business consultancy for self-governing enterprises to obtain more information about their rights. However, their attempt came too late. Dr Schöpf obtained the section and into the bargain an interest-free credit of over 900,000 Deutschmarks. In addition, he was able to take over all machinery, materials and stockpiled machines for one Deutschmark. He was able to rent the production hall for the knockdown price of just ten Deutschmarks per square metre.

At a meeting with the employees in his section on 4 June 1991, he presented them with a choice. They either leave Stanex with a redundancy payment or they could sign a new contract obliging them to work for him. Under this new contract there were no social provisions made for employees should the assembly automation business go under. He announced that he did not want to join the employers' association, that he would pay wages lower than the official negotiated scale rate, and that there would be no 22 percent supplement, should the short time work rule continue. Possibly, so he explained, he even wanted those people to work full time who were officially working short-time hours. After these explanations, Schöpf asked for comments from the workforce. No one volunteered his thoughts. Then he distributed a list of appointments for individual meetings with every single member of the workforce. There was the din of the collective scraping of chairs and everyone fled. It was only when they got outside that they made their anger and frustration known. In their helplessness, many of them cursed the 'Wessis' (West Germans) 'for letting everything here go to the dogs'.

Dr Schöpf's offer posed a dilemma for the employees. They either took the redundancy payment and lost their jobs, or they kept a now uncertain job, together with losing all entitlement to the social fund of the enterprise Stanex. Thirty-two section employees sent a letter to the general manager of Stanex, Mayer, and to the workers' council, and demanded that provisions for redundancy indemnification should be arranged in the event that the new GmbH (Plc) of assembly automation was declared insolvent or employees were made redundant. They received no reply.

The employees of the section assembly automation that Schöpf wanted to keep on tried to ascertain exactly what their rights were according to the agreement between the Treuhand and the purchasing company. However, despite many discussions with the workers' council representatives, they received little in the way of concrete information. Even two days after signing the contract between the enterprise, the Treuhand, and the purchasing company, the workers' council representatives claimed that they had absolutely no knowledge of the content of the contract. As the employees did not receive a written answer to their letter, they had to rely on rumours, word of mouth, and on the often-contradictory information that Dr Schöpf imparted during individual meetings. The workers' council, their official representatives, had in effect been 'bought', as it later transpired that the workers' council representative and his deputy, along with the plant manager

Mayer, became trustees of the job creation company established by the real estate company during the Stanex clear-out.

With neither knowledge of the content of the contracts, nor effective representation of their interests, the employees had to confront opponents in the negotiations who had much experience, the advice of lawyers and well-established connections with the Treuhand. The head of design engineering, Kallmann, summarized the situation:

> The big problem that we have is that all the important people are satisfied with the signing of the contract. Dr Schöpf is satisfied because he gets the assets and machinery for nothing, as well as a loan and cheap rent (DM 10/m²). The combine is satisfied that it no longer has to deal with the Stanex problem. Mayer is satisfied because he could still knock out a pretty decent redundancy payment for the 1,330 employees he had to let go. The Treuhand is satisfied because it has privatized yet another firm and still made 4.5 million. The real estate company is satisfied because they were able to buy a huge plant with new workshops in Berlin, as well as an eight-storey office and production building just 10 minutes from Alexanderplatz for 26 million Deutschmarks and because they were able to get rid of 241 employees in a seemingly reputable way. Because everyone is so happy, it is extraordinarily difficult to be a killjoy. (Kallmann, design engineer, Stanex 11 June 1991)

With the real estate company acting as a silent partner, Dr Schöpf was now the sole leader of a firm little in demand and of whose 62 employees the majority were working on a short-time basis.

The reasons behind Dr Schöpf's wanting to become an entrepreneur, or as Fröhlich put it, 'his own class enemy', were not only linked to economic factors. Dr Schöpf was over the age of fifty and would probably have found it difficult to find a position with a similar level of responsibility elsewhere. He took a considerable financial risk in founding the company, despite his solvent silent partner. He did it to preserve his position of power and to protect the business, his life's work, from the clutches of the class enemy. Fröhlich explained to me what he thought to be Schöpf's motives:

> They believe in God Marx! That's what I think! He was a member of the enterprise militia too. On the weekends they used to take off to the countryside with spades, to practice fighting so that they could defeat the class enemy. Now that's what they've become themselves! You can't just forget that! What's going to happen now? It's inexplicable to me! I mean, the consequence for every single person would have been: 'I have been

mistaken. I have to do something different now. But I can't, for God's sake ...' Yes, the only explanation now is, that they think that they are doing the best for everybody. But that is not what they do, at least not in most cases. (Fröhlich, skilled worker, Stanex, 16 June 1991)

Schöpf created his enclave in a world dominated by market economy. He was able to continue employing his old comrades from the enterprise party group and was not required to yield to a capitalist boss. With the acquisition of exclusive, individual ownership rights he pursued his political-ideological aims. Even during the period of the planned economy, he had taken it upon himself to find loopholes in the system, which allowed him to work relatively independently of the impositions and imperfections the planned economy brought with it. Since his machinery now stood stockpiled and customers wanted to be flattered, he chose an inverse solution to solve his economic dilemma. He explored all possible avenues in order to obtain state support, which would allow him to keep the enterprise. He received short-time work funds, employment incentive jobs (*ABM-Stellen*) for those employees working 100 percent short time and subsidies from the Berlin Senate so that new machinery could be developed. He could also count on the unwavering support of those employees who had been his political allies at the time of the GDR, for example Voigt, his colleague from enterprise militia and the head of production; Karst, the design engineer; and Ruland, the union activist. For them the enterprise was the last bastion against the Western enemy.

Even those employees vehemently opposed to Schöpf, for example Born, signed the new contract. They did not want to lose their jobs until they could take early retirement. Both the employees and the new owner regarded the new ownership structures as political relationships. Through gaining exclusive possession, Dr Schöpf secured his opportunity to continue his dictatorial regime within the enterprise. As Schuster claimed, he put an end to the employees' democratic endeavours, excluding them from enterprise decision making and therefore from taking control of their future careers. He was, however, only able to acquire this right to ownership because the political environment favoured privatization and he was prepared to take an economic risk that none of the employees would.

The acquisition of exclusive and individual rights to their enterprises made neither manager behave in a way that might commonly be believed as proper in the market economy. Instead, private property accorded them so much power that they could afford not to learn

(Deutsch, in Offe, 1994: 11). The specific culture of human interaction common to former socialist economy *Kader* (Dittrich 1992: 324) blocked enterprise development. Thanks to his exclusive right to ownership, the manager of Stanex built a political-ideological enclave in the capitalist world, which he was able to maintain with financial support from the state. Taghell's director saw his opportunity to become rich and took it – without a thought for the enterprise and his long-standing colleagues. In reality, the transfer of exclusive ownership rights to the managers led to the dispossession of employees' democratic rights. The employees of both enterprises were fully prepared to put in time and effort for their enterprises. However, they were overwhelmed by the mysteries of the market economy and hesitated. In the case of Stanex, they were uncertain about becoming owners themselves; in the case of Taghell they hesitated to challenge the ownership rights and the leadership of the managing director. Their collective concern for the enterprise and their colleagues was far more developed than that of the managers.

However, their possibility to democratically influence the privatization process decreased while the West German institutions filled the gaps and the Treuhand became the 'kingmaker'. A combination of the old boys' network within the privatization institutions and a blind belief in the market's self-healing powers destroyed the weak seedlings of enterprise democracy.

Chapter 6

Market Delusion

The high acceptance that the introduction of a market economy found at the beginning of 1990 among the workers was not only because of their consumer appetite, but also because of their need for moral renewal and transparency following the collapse of the 'deceitful' system of the planned economy. The pure market model, derived from the theoretical assumption of a self-regulating system of perfect competition, whereby competing producers satisfy the demand of independent consumers (Preston 1992: 61), received in the enterprises a popular interpretation. It was turned into a 'little tradition' (Tambiah 1970: 3–4), in other words, into a localized version of a large dominant tradition. This had a powerful moralizing dimension. Commercial transactions, the position of the employees in the enterprise and their chances in the job market were judged and justified using moral categories. The employees imagined the market to be a hard but fair system, which would limit the 'inflated social aspect' of the planned economy, reward the competent and hard-working members of society, while punishing the idle and the negligent. The employees had in mind the principle of fair competition, which would be beneficial to society and provide motivation.

In the two enterprises Taghell and Stanex, which were not courted by Western firms, the workers hoped to get rid of the 'inflated administration' and to be able to prove their skills, while the administration believed it would gain a firmer grip on the 'lazy layabouts' in the production (Senghaas-Knobloch 1992: 302) and impose a faster pace. The employees were disappointed and disillusioned when they improved their product and considerably increased their productivity, but failed to secure a foothold in the market, were squeezed out by competitors and let down by suppliers. They recognised that their competitors were using glossy advertising and attractive designs for their product, which was in fact of no better quality than their own.

Their moral objections arose from their experiences of the new economic system as unclear, irrational and arbitrary. Their encounter differed considerably from what neo-liberal ideologues defined as the morality of the market. In the words of Hayek: 'The morals of the market do lead us to benefit others, not by our intending to do so, but by making us act in a manner which, nonetheless, will have just that effect' (Hayek 1988: 81). In their everyday experiences they did not find anything that corresponded to the theoretical assumption that the market mechanism or price mechanism balances supply and demand, informs competing individuals about the market conditions, and has them deal with one another according to economically rational criteria.

Nonetheless, the neo-classical economic paradigm, which reduces human behaviour to a single driving force – the self-interested individual maximising utility and making rational decisions – had spread in the enterprise as a popular explanation for the collapse of the socialist system. A type of economy founded on the self-interest of the individual was interpreted as the 'natural state' of human society, while socialism had forced an unnatural 'social state' upon them. Even the devoted socialist Veit Kater thought, resignedly, that neglecting 'natural' inclinations in humans, who want to have the best for themselves and to compete with others, had contributed to the decline of the socialist system.

> I understood that Marxism only dealt with people as social creatures, as creatures interacting within society. It never dealt with people as biological creatures. That is possibly the reason why the ideas of Marx and Engels failed to work in the end. (Kater, Stanex, 20 February 1993)

The employees who reflected on meaning and justice in the new society evaluated the market economy according to moral criteria that they had developed in the planned economy. They referred to the promise made publicly by the Helmut Kohl that he would transform the GDR into a prosperous country. In spite of all the criticism uttered about the previous economic order, I heard it repeated over and over again that the new system was supposed, 'to build upon what had already been achieved'.

However, the practical obstacles to realizing such an idea were virtually insurmountable. An East German enterprise had numerous difficulties to overcome if it wanted to gain a foothold in the market. The enterprises had too many workers, and machinery that was completely outdated. They had to adapt their product range to suit demand, to seek out markets and learn how to calculate profit and loss. To solve these fundamental problems, the Federal German state attempted to supply credit and grants, finance training courses and provide further education. Where worldviews and personal attitudes, developed from socialist experiences, determine economic behaviour and the view of the market, the self-evident functioning of the market mechanism appears in a new light.

Marketing

The fall of the Wall initially created a vacuum in responsibility for the distribution and commercialization of goods. The state organs of marketing ceased to function or distributed mainly Western products. For the first time, managers of GDR enterprises were confronted with the task of selling their products and adapting to consumer desires. In enterprises like Taghell, which had a solid Western business partner, and in places like Stanex, where much-sought-after investment goods were produced and which had been very successful in the GDR economy, there was cautious optimism. The managers explained they were glad, because the disappearance of the unwieldy state bureaucracy would give them the opportunity to prove their true abilities. The head of process planning at Stanex, Grabher, explained why the department of assembly automation should have been an ideal candidate to survive in the market economy.

> We made a profit of 2.5 or 3 million on our 10 million and the work in process planning and construction played a large part in that too … we always tried to systematize this product, this machine, more and more. And in so doing I can make the product cheaper and cheaper, that's obvious! … In this way we were also to some degree a shining example to the rest – above all in our enterprise, and we didn't do too bad generally in the former GDR economy either. Even if we would have had market economy then, we would have been able to hold our ground. (Grabher, process planning, Stanex, 7 May 1991)

Grabher concluded that the success in rationalising production during the GDR period would portend success in the market economy. He assessed past economic achievements using figures taken from the plan accounts, and consequently made the evaluation using a predominantly fictional form of assessment, which failed to have any bearing in the market economy. Both, the GDR enterprises and the Western investors overestimated the chances of the enterprises in the market economy by using the same calculations.

Up until April 1990, Taghell and Stanex lived in the illusion that they would have their hands full with contracts for the next two years. However, even before the currency reform on 1 July 1990 fears began to grow that their level of production would not be able to keep pace with that of Western firms. In order to prepare themselves for the market economy, the directors began to improve on the weaknesses of their enterprises, which up until now they had kept hidden from the monitoring authorities of the planned economy, namely the low productivity and varying quality of the product. They estimated their chances in the market economy on the basis of problems they had faced in the planned economy and continued to focus their business policies on reconstructing production.

Two of the three enterprises preferred to undertake reform of the production without the intervention of a Western business partner. When Taghell's major customer from Sweden suggested a merger in spring 1990, the managing director refused to engage in this form of close cooperation, which could possibly have undermined his power and control of the enterprise. The head of department at Stanex, Dr Schöpf, reacted in a similar fashion when in 1990 potential buyers began to visit the enterprise. Although a merger or a purchase would possibly have improved the economic situation of the enterprise, it would have also considerably limited the manager's influence. In 1990 he could still operate within a power vacuum, which had been left by the Treuhand to the small and medium-sized enterprises.

Through the exploitation of Treuhand credits and the possibility of registering short-time work,[1] the directors of the two enterprises were able to get by without much financial pressure right up until the end of 1990. They calculated as if they were still under soft budget constraints, which were no longer those of the socialist state, but those of the structural adjustment programme of the Federal German government. Accountants of the main enterprise, which rapidly plunged into bankruptcy during 1991, managed the finances of the automation section at Stanex. The Treuhandanstalt assumed its debts in summer 1991 when it was sold off to a real-estate company. Taghell managed to live on through credit issued by the Treuhand, which ultimately the enterprise could not pay back. Only the third firm, VEBLift, as far back as November 1989, engaged in close cooperation with three Western partners, one of whom later took over the firm.

In the enterprises Taghell and Stanex, who had not found any Western buyers, the willingness for change was more manifest among some of the workers than among the management. The employees of the purchasing departments enthusiastically adopted new materials and half-finished products, thereby cutting back on prep work. Mrs Gertz of materials management at Taghell immediately began ordering materials for lampshades from the West. In spring 1990, prior to the currency reform, the firm Taghell experienced its last boom in demand. The demand from private customers for brass lamps was so high that the enterprise decided at the end of May 1990 to open a lamp shop on the ground floor of the factory building. Mrs Martens, who had a diploma in 'sales point management', was in charge of its layout. In June 1990 the shop flourished, with daily turnovers as high as 11,000 GDR marks. Mrs Martens told me:

> Yes, it was a wonderful time, an interesting time. The customers were grateful that they could get rid of their GDR money. From 1 July, the Western money was introduced. In those days they would just come and say: 'We've still got 800 marks here. Now make something possible and we'll buy it. We don't want to go to the bank. We don't want to pay that in any more because we won't get anything in return.' That's how it was. So we did the craziest things … . We would put a package together and well … I can't even begin to describe it. The customers … they were so grateful they would give us a bottle of spirits. It often came with a bunch of flowers. Well, they were so happy! Word then got around that we could make this and that possible. Yes, that was a great month. (Martens, sales, Taghell, 30 July 1991)

In June 1990 Mrs Martens experienced the final revival of the economy of scarcity. The customers who wished to rid themselves of their last few GDR marks, which they could no longer exchange by benefiting from the profitable rate 1:1, stormed their shop. Just like her colleagues in sales used to, she received gifts if she 'made something possible'. The success of the shop seemed to herald the good times they thought were to follow. Yet, on 1 July the flow of customers suddenly dried up. Mrs Martens was moved down into the sales department where she was now supposed to coordinate the lamp sales.

With unrelenting energy and on her own again, she went full ahead into preparing the Leipzig autumn trade fair. She organized a photographer and had a lamp catalogue laid out, although this was never printed. She made numerous contacts and returned from the fair beaming and full of optimism. She tried hard to promote the name Taghell, launched advertisements in newspapers and lamp catalogues, arranged the Taghell logo to figure on the boards of local football pitches, and obtained small alarm clocks as advertising gifts. Her colleagues in the sales department, who had been there back in the period of the planned economy, let her carry on with her work but gave her no support. In her opinion, even the director failed to recognize the new opportunities she had brought the enterprise. She experimented with advertising campaigns, something that was entirely new to her. She also reverted to the old strategy of the planned economy by using political connections for economic aims. In that way she successfully approached the Berlin Senator of Economics Pieroth at an exhibition and got him interested in the enterprise. She hoped that he would assist in placing the enterprise in the list of lamp manufacturers in the Berlin Brandenburg region, where the department stores sourced their lamps. Yet, the director refused to write a letter to the senator. Even Mrs Gertz, the head of materials management, worked hard in response to Mrs Martens' pleas to gain access to this list via her party comrades from the *Partei Demokratischer Sozialisten*[2] (Party of Democratic Socialists), and was successful. Nonetheless, the director remained unresponsive.

Eventually Mrs Martens got into conflict with the director of the enterprise in her dual function as head both of the workers' council and of sales, and was subsequently made redundant. Her colleagues thought that although Mrs Martens had been a 'nice sweet lady', she had gone too far with the aim of improving sales. None of her successors displayed similar dynamism or were as successful in combining new marketing strategies with older networks. The other employees in the sales department, who from the times of the planned economy were used to being courted by customers, were scared of their clients and no longer knew how to behave towards them.

> I am a bunch of nerves whenever a customer comes in. Today there was again such a situation. First I got a call where he more or less announces his arrival. Fechner [the director; B.M.] likes to shove me in front of him. Next I'm supposed to fool around with the contracts. But he gives nobody the full information. He never says: 'Let's sit around the table, and here's our plan of action and that's what we'll be doing.' – There's never any of that! Some scraps or other are thrown over to us to be digested and then there is an order: 'Do that!' I try to get out of all this; if someone comes along then I just send the customer over to him first. (Brandt, sales, Taghell 31 July 1991)

The employees of the sales department and the director of the enterprise pushed customers between one another and thus also the responsibility for processing the order. The director developed a discount scheme for attracting new customers, which ignored profit considerations. Mrs Brandt regarded it as excessive that she was supposed to negotiate with a customer when she had not received any guidelines.

Mrs Brandt's treatment of clients was arbitrary and hardly accommodating to their needs. When in my presence a customer called and complained that the lamps he had received were more expensive than what was written in the new catalogue, Mrs Brandt awkwardly explained that his lamps had been priced at a previous higher rate and that he would have to pay this price. Soon after, she discovered a contract for a hanging lamp where the exact part number for the corresponding type of lamp cord was missing, and she remarked: 'Now I could be mean and charge the customer for the more expensive lamp cord' – and this was precisely what she did. For the employees in the sales department, the customer seemed to be an opponent, whom they were to fight and swindle, instead of aiming for customer satisfaction. Even when they had the chance to win a good customer, the employees in the sales department were so reserved they appeared hostile.

For the women who were used to being courted by customers with wine and chocolates during the planned economy, the reversal of power relationships between buyers and vendors was unbearable. In the official discourse of socialism, selling and trading goods had the taint of corruption and immorality. In the economy of scarcity employees never had to price and sell their product. The employees of the sales department were now no longer courted, but they had to become the courtiers in order to curry favour with customers. The managing directors of the lamp shops who had previously implored Taghell to sell them lamps, now refused to buy anything from them. 'The people in the sales department are being punished for their earlier arrogance', one colleague remarked. Being forced to sell what the enterprise manufactured was deemed humiliating. This lead to fear of contact also with potential customers from the West. 'I don't want to be involved in any door-to-door selling', was a common objection to selling.

The employees of the sales department waited for someone who 'knew what to do'. However, there was a new head of sales every three months and consequently the sales strategy kept changing. Initially, there were the advertising campaigns of Mrs Martens, but these were brought to a standstill. A computer was purchased in order to run a customer database, but this was never put to use. Finally, the lack of ideas coming from the director allowed the new head of sales, Bürger, to buy up a large supply of lamp elements from a West German firm, although Taghell had no use for them. A persistent rumour circulated that he had earned a hefty commission for himself in making this purchase.

A glance at the catalogue which Taghell finally commissioned for its customers, reveals in the first few pages a mixed collection of cheap lamps which were bought in addition to the Taghell production: desktop lamps with a pewter base, a glass kitchen lamp with small golden bows, standing lamps with turned wooden bases. In the sales representative's opinion, customers were not interested in any of these lamps. The head of sales, Bürger, had added them to the collection, in order to 'lighten' it, in his words. The brass lamps designed and produced by Taghell appeared on the back pages, hidden away between lamps made from the parts that had simply been bought in. The catalogue not only reflected the lack of experience but also the lack of trust that the employees had in the value and the quality of their own products.

At Stanex too the marketing strategies were developed against the will of the department director. In spring 1990, two employees went searching for customers on their own initiative. While the department director Dr Schöpf was on a spa holiday for five weeks, the engineers of project planning, Born and Schadorf, tried to explore the West Berlin market.

We did that on our own private initiative. My friend in West Berlin, who has been self-employed for thirty years, told me: 'You two should get yourselves to the department of industry and trade, there you'll find a sort of fat volume [directory of businesses in the area; B.M.]. In there you can make a few enquiries. It's all in there.' And straight away on that very day, we set off and started collecting addresses for the customer database. We wrote letters. Then finally we presented them to our general director here, to Mayer, who then wanted to make two or three changes, which were incorporated into the letters. Then he said: 'Right. I'll sign this as the managing director so they can be sent off.' Then we copied them for the first hundred companies, well we had someone write them for us. A typist, who did not even belong to us, but was from a completely different department, did that for us out of kindness. But when he returned from his spa holiday, he [director of the section automation B.M.] Dr Schöpf said: 'We aren't going to do that, these letters aren't going anywhere!' And as a result we had to tear up the hundred and then there was calm again. To this very day not a single letter has been sent off. (Born, project planning, Stanex, 4 February 1991)

Since they lacked even the most basic advice of how to go about things in the market economy, the two employees turned to an old friend who worked and lived in the West. His tip to go to the department of industry and trade (*Industrie und Handelskammer*) and to make a list of potential customers from the business directory seems unoriginal, but it was the most concrete piece of advice that they could get at the time. The market was a completely new system for them and had rules they did not master. They knew enough, however, to take action immediately to prevent themselves from going under along with their firm. While the general director Mayer supported their initiative, the director of the section automation Dr Schöpf resisted any initiatives if they were not originally his own.

In summer 1990 he forbade them from searching for customers in West Germany or to visit the plant of their competitors in Baden-Württemberg. Two months before the official unification of both German states, he argued that West Germany was Western foreign territory and his employees were not allowed to travel over there because they had not acquired the political status of a travelling cadre (*Reisekader*) in the socialist system. Dr Schöpf clung to the old political division of the world and wanted his product to be of benefit only to customers in the socialist block. However, from May 1990 the East German electric and electronic industry, their main customer, had cancelled nearly all of its orders and since summer 1990 had been in such a critical economic situation that it could not take any further investments into consideration.

> At this point Dr Schöpf put his foot down and, well, forced us to focus on the former GDR enterprises. But it was clear to all of us at that time that nothing would come of it because it [the economic collapse; B.M.] was practically round the corner ... so at times we were actually standing in front of factory gates which were chained up and locked. There wasn't even a gateman on duty anymore. (Mahler, head of project planning, Stanex, 2 May 1991)

Dr Schöpf had always perceived the creation of the department for assembly automation, in the sense of the official ideology, as a struggle against Western imperialism. His machinery was to strengthen the economy and therefore also the political system of the GDR. Consequently, it was now inconceivable for him to offer this to Western customers. This conduct was hence in complete accordance with the views he previously expressed as the absolute truth for years.

The few East German customers who were still doing business had not forgotten the unacceptable customer service and the draconian conditions that Dr Schöpf had exposed them to and were not interested in working with him any longer. The construction engineer Scheuch recalls:

> Dr Schöpf then decided that several groups, mainly people from project planning, travel throughout the GDR to get into contact with former business partners. Some of them said to us: 'As long as Schöpf goes on doing that, we won't be buying anything.' They all knew Schöpf and many didn't really take to him at all. (Scheuch, construction engineer, Stanex, 24 July 1991)

The director of the section automation remained caught up within his 'mental geography' (Darnton 1991) right up until November 1990. When he finally declared exploring the Western market the priority, the economic situation of the enterprise was extremely precarious. Potential customers were reluctant to take the risk of investing half a million marks in a machine produced by an enterprise that might go bankrupt at any time and that could not guarantee maintenance and delivery of spare parts in the long term. Grabher, the head of process planning, remarked:

> 'In January [1991; B.M.] Schöpf stood up and said: 'Right, we are half a year behind on work in the market.' At which point I told him: 'It's your fault that we are now half a year behind in the Western market. Last year you called us over to the East when we were trying to get started with selling our products in the West. You had all of us sit around, instead of allowing people – which of course costs money – to travel around West Germany so that the name of Stanex establishes itself.' (Grabher, process planning, Stanex, 7 May 1991)

Once his hopes for the East European market had been completely shattered and he had failed to attract any business, in autumn 1990 the director of the section automation agreed to send letters to potential customers in the West introducing his product. Yet, the letters were still written in the old style of command, which producers could adopt towards consumers in the economy of scarcity. This is how he presented his offer to the customer: 'By means of this offer we give you an idea about the technical and economic use of assembly machinery and we would like to orientate you towards a further collaborative problem-solving process' and in the same vein: 'In this process the stated offers are to be developed to the stage of laying out the basis for a contract.'

Unperturbed by the fact that in the market economy the customer has the possibility of choosing from a variety of suppliers, Dr Schöpf continued to formulate his letters as if he was the only producer of assembly machinery and his offer would automatically win a contract. He refused to acknowledge what was accurately expressed by one of his colleagues: 'We used to be the only gods, now there are lots of them.'

> The customers we had in the West at the time, you know, in 1986 … Who would talk about them today? Time flies by so fast! So many firms developed in the market means of rationalisation because it was a branch of business that had been neglected also over there … To get into the market again you need months and years of intensive market research, involving presence at trade fairs, doing advertising and presentations again and again. We no longer have so much staying power! (Grabher, process planning, Stanex, 7 May 1991)

The employees in process planning complained about the single-handed effort of their director and felt they 'had less rights and were more oppressed' than before. It was not until spring 1991 that Dr Schöpf mobilized all the engineers he could muster to search for customers in West Germany. People asked to make them an offer, but a long time passed without any definitive contract. Since the design engineers were not allowed to speak to customers directly, they received technical details about the product for which the machine was to be developed only from Dr Schöpf, who got involved in all the details of the process. Nonetheless, the enterprise received some requests from interested customers, although only one actually resulted in an order for a roundtable assembler. In summer 1991, when 80 percent of the workforce was already working short time and the director had become the owner of the company, he finally acquiesced to taking on smaller and unconventional orders.

The difficulties faced by director Schöpf demonstrate to the extreme the problems of starting to think in terms of the market economy. The director's priority was optimizing production. Marketing and advertising took second place. He leased expensive CNC-lathes and milling cutters, although the enterprise had no contracts to work on. He refused to invest money in catalogues or business trips. He was convinced deep down that consumers ought to be grateful to the producers.

The stance of the employees varied between pride along with a strong identification with their product and lack of confidence, especially towards Western customers. The vendors felt as if they were selling themselves along with their product. They felt judged by the potential customers according to whether their product succeeded or failed to please. The relationship between East German producers who could not sell their products, and West German consumers who did not want to buy them, was considered by many as symptomatic of the German–German relations following the Wende. After the initial months of enthusiasm following the fall of the Wall, when West German companies were still welcoming East German manufacturers, routine settled in and East German representatives were brushed off when their product was not interesting. Mahler, the head of project planning at Stanex, described the painful experiences he had gone through during this period:

> In the beginning there was a relatively good feeling, everything was developing positively as we tried to stimulate interest over there in our product. We noticed there was still some sense of euphoria. They encouraged us: 'Of course, together we'll all get by.' But already by December things were beginning to look different. Perhaps they realized we were potential competitors and after all we were in fact capable of something and weren't merely to be pitied. And it looks now as if they are getting a little hostile towards us … It has even happened that when we went to a business partner, they told us we should stop bothering them, because they had to work and earn money to feed us as well. (Mahler, head of project planning, Stanex, 2 May 1991)

The East–West German prejudices influenced the way the sales situation was interpreted. The East German vendor felt he had to defend himself against the image of an almsman, or in the words of another vendor, against the perception of being a beggar. At the same time, he assumed that the West German customer feared him as a potential competitor and rejected him for that reason. Concluding from the unsuccessful sales situation that West Germans purposely kept East Germans in the position of second-class citizens because they feared them as competitors, helped to restore his self-esteem. This ambivalence was perceptible among all the employees who were trying to find customers in the West. Some argued excitedly that they did not want to sell their ailing product in the West, only to stress minutes later that Westerners did not buy their quality product on purpose.

Most vendors agreed that during the first few months after the fall of the Wall, East German manufacturers could benefit from the psychological advantage of the curiosity and emotional commitment of their West German business partners. This emotional aspect vanished when the costs and problems of unification gradually surfaced and the blooming of the East German economy was projected into the distant future.

In reverse, among East Germans a process of disillusionment with the 'Westerners, as slippery as eels' started to take root. I heard time and time again at Taghell about the alleged poor quality of the pre-polished brass from the West, which had made them shut down their own fine grinder. There was a rift between the perception and reality of the market. The construction engineer Veit Kater explained the difference between the theoretical study of the 'laws' of the market and the practice:

> Not one of us had experience with something like it because there is a huge difference between this type of market which we were used to, which consisted of finding someone who could provide goods, and this other type of market which we have now, which you yourself have always known and that consists of finding someone who wants those goods. And this leap into the new form of market, even if you knew about that already, by reading about it or attending a course for half a year about foreign trade, this knowledge from books is one side of it all. But actually putting that into practice in the real world, that's a tremendous difference. That's where a lot of other people have made mistakes. (Kater, construction engineer, Stanex, 20 February 1993)

The practice of buying and selling in the years after the fall of the Wall was shaped by emotions, perceptions and personal differences, which cannot be found in any textbook. The reversal of the relation between buyer and supplier, and feelings of insecurity and opacity, shaped the perception of market economics.

Working for the Market

The absence of transparency in market relations persisted inside the enterprise, where the old management fought for its position and kept employees in the dark about its decisions, information and strategies. For postcommunist power holders who wanted to hold on to power in an uncertain situation, it was inappropriate to grant workers the chance to understand the situation they were in or to demand their rights by means of clearly defined rules. This mechanism was typical for the postcommunist powers-that-be, whose power is 'too weak to share and to delegate it, by investing it in rule making' (Elster et al. 1998: 33).

Due to their lack of information the employees resorted to speculation and to gauging the economic state of the factory and their own status within it from the trivial details of everyday life. They observed that there was no more toilet paper in the factory toilets and concluded that management had become completely indifferent to the comfort of their employees. They discussed which jobs the cables lying in the shelves were meant for and wondered whether it would be big enough in order to be produced at a profit.

Rumours persisted at Taghell that the department of prefabrication was going to be closed and all the machinery scrapped. Due to changes in the purchasing strategies some work processes were indeed redundant. The workers in prefabrication noted exactly how many new brass lamps were introduced into the assortment, for which the components were purchased in the West. The first brass tubes from the West released a minor technological revolution in the factory. Since the new tubes had already been ground, the grinding department was closed down.

When the director was confirmed in autumn 1990 as the 'new' managing director, he began to dismiss staff on a grand scale and called off the 'pact for plan fulfilment'. The foremen were ordered to increase production norms by 15 percent and to punish serious mistakes and the production of rejects with formal warnings. The foremen themselves received formal warnings if the staff they supervised infringed the rules. The director attempted to reduce the number of staff in prefabrication as quickly as possible. As many of them had worked in the enterprise for most of their lives, he tried to skip the long periods of notice of dismissal they were entitled to. The director avoided contact with the workforce as much as possible; the director of production too could scarcely be seen anymore on the shop floor. The fitter Friedemann commented:

> Now they go right out of their way to avoid us. They may peer into the workshop perhaps. 'Right, there's a few left in there,' but they would hardly ever come in and say 'Hello' or simply come and talk to us. They don't consider it necessary. Otherwise you could complain to them about how bad it all is. They avoid that at all costs. (Friedemann, fitter, Taghell, 29 July 1991)

However, the staff capacity had been reduced so much in spring 1991 that the fitters had to work in production. The fitter Friedemann, who due to domestic problems did not have the strength to look for a new job, described the factory in July 1991 as a place of deliberate destruction, where there was a shortage of absolutely everything, even soap and toilet paper. After the new orders from Sweden in summer 1991, the prefabrication was under intense pressure, but largely impoverished. The workers, who were convinced that it would only be a matter of time before they lost their jobs, took up old forms of resistance from the days of the plan fulfilment pact. They refused to work and turned a blind eye to the disciplinary instructions issued by the management.

> I have to say in all honesty, we used to drink our coffee early at half six or during the day. I'm not changing that routine, why should I? That's going to remain the same. (Friedemann, fitter, Stanex, 29 July 1991)

The women in the assembly department identified more strongly with the enterprise and the product than their colleagues in prefabrication. Since they were left in the dark by management about their business situation, they tried to develop among themselves an idea of their chances of success in the market. They told each other about visits to West Berlin, where they explored goods on sale in the local lamp shops and compared them with their own products. They were willing to 'tighten their belt' and work harder in order to ensure the survival of the enterprise. In autumn 1990 they increased their work performance by as much as 140 to 145 percent of the norm. Instead of starting at half past six, some women were already there by six o'clock in order to achieve their daily quota. When the lay-offs continued and the norm was raised yet again, in spring 1991 the women returned to a collective control of their work performance and to 115 percent norm fulfilment.

The women did not have the chance to compare their situation with others in the market economy. Their analyses of the economic situation of the enterprise were based upon everyday experiences and rumours about the director's intentions. They could still act collectively and either refuse to sustain their work performance or raise it. In July 1991 the women reacted coolly to a major order for more than 25 thousand five-armed and 9 thousand three-armed brass lamps from the Swedish client, which was supposed to be ready by the end of October in time for the busy Christmas period. To achieve the target of producing 380 lamps a day – an achievement which was hardly feasible with only thirteen women and an output of 238 lamps per day – the director of production demanded a 10 percent rise of the norms and announced that employees would be transferred from administration to the factory floor to help out. The foreman emphasized that thanks to intervention the rise in the norm was now a mere 5 percent. The women were unimpressed and made it clear that they found it unacceptable for white-collar workers to be transferred from administration to the factory floor while three young workers were to remain on short-time work. Some female workers put forward proposals to work extra shifts or to introduce a division of labour, as on an assembly line. However, these ideas were rejected with indignation by most of their colleagues. A worker who had been a shop steward for many years explained the point of view of the women following the meeting:

> It's absurd. By and large, there are women who have been working in this job for over thirty, thirty-five years. We don't achieve quite as much as perhaps the young ones do, you know, but as we are finding out, the young crowd doesn't manage all that much any more either. They don't achieve much at all. So we'll have to spread out the workload a bit. That's the right way to do it. You can't leave work exhausted every evening, then you won't be able to do much the next day. (Schmidt, assembly worker, Taghell, 31 July 1991)

The women resorted to 'spreading out the workload' and refrained from overexerting themselves for an enterprise where they did not feel respected as a workforce.

At Stanex as from winter 1990 the head of production Voigt placed greater value on maintaining Tayloristic virtues such as punctuality, order and respect towards superiors. Since the formal hierarchy in manufacturing did not correspond to a hierarchy of experience and qualification, the workers continued to confer directly with the engineers and not to consult their foremen about production problems. In spring 1991 they succeeded together with the construction engineers in developing and manufacturing a machine in only three months, which would have taken them a whole year before the fall of the Wall.

Speeding up the turnover of orders failed to improve the economic situation, as Stanex did not receive any orders in spring 1991. The company only survived because the Federal Government extended the regulation for short-time work in East German enterprises. To a large extent, the workers were paid via the Office of Labour and earned about 70 percent of their East German wage, which in turn was about 60 percent of the West German wage. When the short-time work regulation threatened to run out, Dr Schöpf, who had become the owner in summer 1991, chose the employees he wished to keep. His choice was based less on their qualifications than on their political orientation. His comrades from the enterprise militia were allowed to stay on, while skilled workers and engineers were designated for dismissal. A 'rationalization' of leadership style that Aderhold et al. found (1994: 125) among managers in other privatized East German enterprises, did not occur in the case of Dr Schöpf. His leadership style remained autocratic, without predictability or transparency. Although the majority of employees deemed selection according to the old political criteria unfair, they did not voice their anger publicly. The managing director of this ghost firm, by becoming the master of the workplace, was now the unrestrained ruler over the enterprise.

> Nobody dares to challenge anything anymore. If you think about it, that's logical in such a situation! Everyone is afraid of losing their job, everyone fears the future, everyone hopes that if they don't stand out too much from the rest they get a chance to continue working here. That's if the place survives – that's the main issue, if we subsist and belong to the lads, stay in the circle of those who'll have the privilege to go on working here. For lay-offs are imminent unless there is a miracle, an incredible reversal in the work situation that we need even more people than we had planned. (Ruland, skilled worker, Stanex, 18 April 1991)

Fear of unemployment also granted Voigt, the head of production, an unexpected rise in the power. Many workers, though by no means all, would no longer dare to contradict him openly, even if they thought him totally incompetent. Voigt legitimized his authority not through professional competence, but solely on account of the new authority to instruct and to sanction that the Wende gave him (Aderhold et al. 1994: 109).

> A manager in the market economy has a smoother ride than one in the planned economy. Work in the market economy is an issue of survival. In the planned economy the livelihood of every single person is secure. In the market economy, it is easier for a manager to be in charge, to accomplish his tasks and to impose himself. In the planned economy, we were too social. When problems arose: work discipline, discipline in the workplace, attitude to work, a state-controlled leader was limited in what he could do in the planned economy. (Voigt, head of production, Stanex, 5 February 1991)

The employees were afraid of him but they did not respect him. Horst Fröhlich continued to portray Voigt as an 'obedient control organ' of Dr Schöpf and quarrelled with him. In his opinion, 'a good manager motivates staff with his performance and charisma, but not with a whip'.

Fröhlich was irreplaceable because of his special know-how in the construction of assembly stations. He was the only one who worked almost continuously even in the years that followed, while his colleagues suffered from the uncertainty of working short time or left the enterprise altogether. The power formerly wielded by the employees in their position of irreplaceable production workers was now crushed. Although their short-time work had been reduced to zero working hours, many continued to come to work in the following months as a form of active therapy to counter their depression and to stay in practice. For the majority, the ailing enterprise was a 'cock of hay and they were simply clutching at straws'.

Picture of the Prevailing Mood after Two Years of Short-time Work

An afternoon in the workshop at Stanex in January 1993 gives an idea of the prevailing mood after two years of short-time work.

Fröhlich and Ritter were sitting alone in the manufacture of assembly stations, where eleven workers had been busy in 1989. Four more workers were occupied in the other sections. A factory meeting had been arranged for 2 p.m., during which Dr Schöpf wanted to comment yet again on the current situation. Their colleague Ludwig came by to attend the meeting, shouting as he entered the shop that he was fed up with shopping and doing the laundry. Schöpf should admit that everything was over, then everyone would know where they were at. Before the meeting got underway, Ludwig quickly sharpened a drill bit he had brought from home. He told me that people were fond of each other here despite the 'appalling situation'. Then he told me about the Mazda car which had been stolen from his daughter-in-law and was found in Usedom. The colleagues discussed the best method of protecting their car and Ludwig offered Fröhlich a fuse switch for his Wartburg car. Then he told everyone about the new lounge suite that his son had bought.

Fröhlich pointed out to me that of the six people working some were in reality working short-time and had come to work because they could not stand it at home. He found it difficult to be the only one fully employed because he had to see the suffering and demoralization caused by the lack of work. At home these worries did not get to him so much. All the same, the tense situation had already led to rows within his family. Since December he no longer spoke to his father-in-law in West Berlin.

When they described their problems to him, he had felt provoked and had told them to leave his house if they only wanted to complain.

In the final assembly two fitters were still working. Other colleagues who were heading to the meeting gradually joined us. The fitter, who had come from the Stasi[3] in 1989, was also present. The colleagues discussed attacks and burglaries and thought up brutal punishments to take revenge on the burglars. One of them suggested breaking their spine with an iron bar. Another worker, whose summerhouse had been burgled, boasted he would claim from his insurance a new colour television set, which he had never even owned.

Finally Ruland came and told me how depressed he felt because everything had turned out even worse than expected. He saw it as a positive sign that a lot of people had turned up for the demonstration to commemorate the murder of Karl Liebknecht and Rosa Luxemburg. He estimated the number of participants at 180 thousand and regarded the official figure of 30–40 thousand that was circulated in the media as pure misinformation and lies. People are fed more lies these days than before in the GDR regime, he claimed. At the moment, he was living off his short-time wage of DM 1,100, his wife received DM 280 dole money. Only his children 'were doing well' and had good jobs. He emphasized that he did not believe socialism was dead, because this capitalism was no solution.

Private ownership, its augmentation and the security it provided, formed an integral part of the conversations among the short-time workers. The pipe dreams the men developed were full of aggression against those who could take something from them. Private ownership had indeed become the focal point of their lives that work no longer filled with meaning but with anxiety. Also, what the West German state offered them in compensation for the loss of their work was money for consumption, as they were too old to apply for professional training.

Consumption

Workers and employees followed similar strategies of expenditure that they had otherwise criticized in their clients. In 1990 they preferred to spend the highly desired (West) German mark on Western products. On the one hand, they worried about how they would manage on their low wages or on the short-time wage, when the rents would rise as announced. On the other hand, they purchased in the first three years after the Wende all the consumer goods they had longed for: a video recorders, colour televisions, new sitting-room furniture and often a new car as well.

Their strategies were similar to those of Mrs Schmidt from the assembly department at Taghell. She introduced the most stringent measures to save money at home and limited the amount of fixed monthly spending to be able to face difficult times ahead. She also indulged in big expenses for consumer goods because she enjoyed it.

> No, well no, nothing's changed at our place. The only thing is: three years ago, no two and a half years ago, we bought new furniture for our flat when my daughter moved out. We had no choice. New bedroom and living room. Then just the other day, I said to my husband: 'I really can't stand this lounge suite any more.' They're only two and a half years old. So we drove off the following Saturday to have a look at some new ones. My daughter will take the old ones. She will be real happy when she gets them. That's not something you could throw away yet.
>
> I would never have done that before. I would have kept the old furniture for longer and thought, well, so what? But now, because you see something nicer you buy it, without thinking whether you really need it or not. (Schmidt, lamp assembly, Taghell, 31 July 1991)

She refurbished her flat and gave the old fittings to her daughter. Of all the people I interviewed and whose homes I visited, everyone had done up their living room. Summerhouses were crammed with things they had thrown out, such as pieces of wall-to-wall cupboard units and old couches. At the same time, the women discussed at work for hours about where they could get the best value for everything and how they could cut costs on electricity and gas, which became more and more expensive. Mrs Schmidt told me that she no longer made fruit and vegetable preserves because gas prices had risen so high.

> Life is really expensive these days. If I think about it, boiling or preserving food takes a few hours. Even if I just think about the price of the gas. It'd be cheaper for me to buy it all ready-made. (Schmidt, lamp assembly, Taghell, 31 July 1991)

There was also the reaction of astonishment and disbelief that more or less everything can be purchased in consumer society. After a walk through the red-light district in Amsterdam, the engineer Scheuch reflected upon the apparent boundlessness of consumerism and came to the conclusion that people who could afford everything must feel desperately unhappy, almost like dead:

> But that's when I noticed how distant I had really been from all of this. I really felt quite shocked deep inside when I saw all of that with my own eyes. What consequences the irresistible urge to make money can have. When there isn't anything any more that isn't there already. And what I find particularly bad is that the rich who can buy everything, they probably can't enjoy anything anymore. What if I could afford everything?! I try to picture what it must be like. I don't know what it's like to be a millionaire: Nothing ever runs out. I sense that it keeps growing in bank accounts with interest. But who can say whether that's a nice feeling or not …? (Scheuch, construction engineer, Stanex, 24 June 1991)

After forty years of the planned economy, where he had constantly refused to defend the illusion of the plan, Karl Scheuch desired a moral renewal of his society. He expected that the market economy would operate without false pretences and according to objective laws. When this image of the market was shattered, his role as consumer became ambivalent. Like many other East Germans he discovered that the joys of consumerism were on no account as great as he had hoped and expected. He learned what Hirschman (1984: 46) described: always, whenever economic progress opened access to consumer goods for certain social classes, intense feelings of disappointment or hostility were stirred up towards the new material wealth. After his initial enthusiasm for the pleasures and possibilities of consumer society, Scheuch became suspicious that the new society had failed to keep its promise.

The employees at Stanex thought a great deal about the blessings of the consumer society. Veit Kater wondered whether his quality of life had really improved once he could buy, within a period of three years, all the consumer goods that GDR citizens used to dream about for all their lives and which sometimes were a novelty even for West Germans.

> Well, three years ago I still didn't have a car. Shortly before the Wende I bought a colour television, my first one. I didn't have a video recorder back then, although I've got one now. Neither did I have a hifi system and my camcorder and so on … basically, I'm better off than before. But whether everybody is really aware of the disadvantages that go hand in hand with this lifestyle? Whether it is clear, what 'quality of life' actually means? (Kater, construction engineer, Stanex, 20 February 1993)

His colleague Horst Fröhlich gave a pessimistic response to this question in the form of a saying that is frequently cited by environmentalists, and which has been attributed to a Native American:

In fact, this can only end like the saying goes from a North American chief: 'People will only realize that money isn't edible, when the last river has run dry, the last tree has been chopped down and the last bird has been killed.' That's the gist of what he said somewhere. That's what is happening, I'm afraid. Yet, I don't really know what I can do to prevent it happening, other than to sit around and suffer from knowing that it will happen! (Fröhlich, Stanex, 16 June 1991)

The active and enterprising employees at Stanex and Taghell regarded themselves as having a double handicap. They did not master the conditions of market economy, though they were willing to experiment, and their attempts at finding creative solutions were crushed by company management.

The uncertainty about their jobs destroyed their identification with the enterprise and the way they viewed themselves. As working individuals they became exchangeable. Instead of being wooed members of the working class, whose 'personality' was in demand, they were now merely part of a workforce, whose jobs and enterprise were constantly under threat. Control over the material conditions of life slipped from their grasp. In the new, changing, political-economic context, the market proved to be a battleground of power relationships (Dilley 1992: 4), in which workers and employees were the losers. Some of the people I spoke to went so far as to describe the economic collapse of their enterprise as a 'death'. One foreman, whose department faced closure, said: 'it is being killed off'. Whenever a worker faced redundancy, his colleagues remarked: 'He probably won't survive.' When reflected through the prism of the power structures in their enterprise, the market economy appeared to most of the employees of Taghell and Stanex as arbitrary, irrational and violent.

Notes

1. The government programme for short time work was helping enterprises in financial difficulties to continue employing their workforce in periods of low demand. The enterprise paid the workers part-time, while the state provided the funds for paying workers full time.
2. The Partei Demokratischer Sozialisten, PDS, was the successor party to the SED.
3. *Stasi* from *Staatssicherheit*, which was the organization of secret police in the former GDR.

Chapter 7

Worldviews at the Time of the Wende

In the time leading up to the Wende of autumn 1989, any promise for the future lay in the emphasis on common ground between East and West. With exhilaration, the cry of 'We are one people!' was to be heard in East and West. As the Wall came down, thousands of West and East Germans fell weeping into each other's arms. We watched these scenes of emotion between complete strangers on television, and hurried to be at the Wall ourselves. But, since then, that feeling of unity has been lost. You could almost think that it had never existed, and that the essentialising discourse of unity had never been this irresistible tide that had carried the German reunification along with it, against all the doubts and worries of the neighbouring countries.

In the Berlin of 1991, the differences between *Ossis* – East Germans – and *Wessis* – West Germans – were an integral part of every pub conversation. It became a sport to try and pinpoint the origin of the people on neighbouring tables, by looking at their clothes (although this became increasingly less obvious) and also their way of speaking, moving or laughing. In most discussions and comments on the Wende, employees of East German firms accepted that there were fundamental differences between East and West and highlighted their specifically East German identity. Over and over again, I heard the following assertion, which Anderson (1983) considered to be a constitutive element of 'suppositious commonality': my East German interviewees claimed they could immediately tell if a colleague was East German, even if he was working in a Western firm, simply from his way of speaking and his manner. Some East Germans began to define themselves in contrast to the picture that they had created of West Germans. Dialogue between East and West became visibly more emotional and pervaded with prejudices. At the end of 1991 I heard West Germans characterised as being socially isolated, obsessed with work, unable to share and indifferent towards the developments in the former GDR.

Self-reflexive tendencies in social anthropology in recent years (Rabinow 1986; Cohen 1994) have sharpened the awareness that cultural representations such as worldviews and self-portrayals are irrevocably connected to power. The idea of rivalry between cultures, such as, for example, between that of East and West Germany, or of the dominance of one culture over another, has given the concept of culture itself a totalising aspect, which is used in countless power struggles on both micro- and macro levels. Although 'culture' has seemingly entered everyday speech without problems, as a concept it has a long and turbulent history. In social anthropology the term is used, above all, in two different ways.

First of all, a differentiating concept of culture exists, which involves attributing characteristics to a particular group of people, characteristics that shape their social behaviour and their point of view on society. 'Culture' serves to define 'the other', based on the hypothesis that the culture of a people, its ethnicity, or its nationality, constitutes its essence. In other words, this means that people do what they do, because they are what they are (Friedman 1994: 72). To remain with our example: the *Wessis* ' are letting everything go to rack and ruin here', simply because they are *Wessis*.

On the other hand, to have ' culture' is a human quality, which organizes behaviour according to patterns of meaning, in contrast to simple reactions and instincts. It expresses the ability of people to develop plans and models, which guide their intentions and influence their paths through life. It is a characteristic that distinguishes humans from biologically determined species (Friedman 1994: 72). This second meaning is dominant in British social anthropology, which supports an idea of culture as everything which is, or which can be, learned, and which, therefore, can also change.

West German politicians, management consultants, and educators contradicted the essentialising approach and based their requirements for the adaptation of East Germans to the Western way of living and working on this second concept of culture, which presupposed that 'everything can be learnt'. Training programmes, financed by the Federal Government, were aimed at balancing out the differences between East and West, as quickly as possible, through education.

However, just as the debates about a West and East German culture show, ' culture' has already become a concept that has been appropriated by those involved, with much political significance. They use the term in a variety of ways, according to their political and economic interests. I would like to venture into the lion's den, as Herzfeld suggests, and take as my subject the stereotypical discussions about the culture of the 'other', analysing them as an instrument for following up interests and strategies, from which even the author herself is not free (Herzfeld 1997: 67).

Although East–West stereotypes are ubiquitous, opinions as to what is typically East or West German differ considerably, and, significantly, the East German can feel that something is typically East German that the West German sees as being typically West German.[1] Differences and similarities in lifestyle, in beliefs, worldviews and values cut across the divide between East and West almost systematically. So what is there still to investigate, if the subject matter apparently seems to be dissolving?

The Construction of Two German Cultures

Walter Ulbricht spoke at the 7th Conference of the Socialist Unity Party (SED) in April 1967 of the development of a socialist national culture, which had apparently led to the coexistence of two rival German cultures, irreconcilably pitted against each other (Meuschel 1992: 216, 400). But had the GDR citizens internalised this official interpretation? Could it in fact be traced back to forty years of real existing socialism, to the two different histories, or had it only gained relevance after 1989 because of political and economic experiences, as a micro-political act of differentiation against the dominant West Germans, against relatives, employers and bureaucrats? Is it something that one can fall back on, in order to defend economic interests and as a source of cultural and psychological security (Friedman 1994: 76)?[2]

The East and West German stereotypes that I encountered in the firms were nurtured by real experiences, for example, relationships with West German relatives, the experience of selling goods in the West, the contact with and exposure to Western institutions, and so on. I shall investigate how they function and which real power relations they conceal, and make use of.

Western Kin

Many of my interviewees in East Berlin firms had relations or acquaintances in West Berlin or West Germany, who had been their links with the West before the Wende, even though contact with them was only sporadic, limited to parcels at Christmas or postcards from holidays. This direct personal contact certainly contributed as much as Western television to the fact that, before the Wende, an image of the West had been created that did not have much in common with reality. Roles in these relationships were, for the most part, clearly allocated. East Germans were expected to complain about their material situation and be grateful for gifts from the West, while the West Germans were supposed to boast of their success and their material achievements. Thus, to be permitted to travel to family occasions in the West was always a point of confrontation with the regime. It had far greater significance than merely a family matter. It was a chance to adjust one's view of the world and to compare discourse and ideas against direct observations. Nevertheless, it was only after the fall of the Wall that many East Germans were able to verify, with their own eyes, the degree of truth in the stories of their Western kin, and see more of West German reality than prettily decorated coffee tables in tidy apartments or smart family homes. Presents from the West had been a means of distinguishing oneself from socialist uniformity; in many East German kitchens that I visited there were still, two years after the Wende, empty coffee jars and wine bottles from the West standing as ornaments on tops of kitchen cupboards.

After the Wende, it was a huge relief for the foreman of the prefabrication department at Taghell, Franz Saller, to no longer have to be grateful to his Western relatives. At a family gathering he made it clear to his aunt in West Berlin that he now understood that the gifts that she had brought him were not actually all that valuable.

> Like I said to my aunt in Reinickendorf, West Berlin, when we visited her recently for her birthday, 'Actually, you know, I wanted to get you a pound of coffee, like you always used to get us. But that would've been really embarrassing for me, to buy you such a cheap gift.' (Saller, foreman, Taghell, 21 January 1991)

The aunt took his observation, which in many other German families would have led to lasting tension, with good humour. ' She laughed, she actually laughed.' Saller's wife, however, reprimanded him in my presence and reminded him, 'Come on, and don't forget that before coffee would've cost us 35 marks.' And then he felt obliged to remember his earlier gratitude, which he had so eagerly abandoned: 'We were always pleased.' His wife confirmed this: 'Of course we were always pleased when we got a packet of coffee. Saved us 35 marks.'

Franz Saller was trying, with his provocative behaviour, to counteract a stereotype that he felt trapped within: that of the 'grateful East German relative'. Before the Wende, the stereotype of the East German as the inferior 'taker' of gifts had dominated. In accepting supposedly valuable gifts they fell into a state of dependency towards the West German 'giver' (Mauss 1968: 146–47). Of course, East Germans also gave their Western friends presents, which usually cost them more, both financially and in terms of time, than the West Germans spent on their presents. Saller described to me at length, the amount of trouble he and his wife had gone to, in order to buy and send a porcelain dinner service for a wedding in the West. After the Wende they were able to admire the same expensive service for a quarter of the price at the Karstadt department store.

After the currency union this inequality of giving became a thing of the past. With his remarks, Saller was satisfying a new stereotype that had developed since the Wende: that of the Ossi who makes many demands and is ungrateful. This new stereotype was not unknown to Saller, but by consciously taking on this role of someone who recognises the value of things and who can also be ungrateful, he turned the negative stereotype into an expression of self-confidence.

While many of my interviewees were freeing themselves from dependence on their Western relatives, they were reversing their political criticism. While they had earlier given free rein to their criticism of the political regime in the GDR, after the fall of the Wall they shocked their Western contacts by stressing, 'it wasn't all that bad then'. Above all, many Western relatives frowned upon criticism of the politics of the new Federal Republic. Karl Scheuch, a former design engineer at Stanex, told me about the behaviour of his sister-in-law, which had deeply shocked him.

> I've got a brother who's a lawyer in West Berlin. Well, his wife is training to be a teacher and she's got a terrific job with the government. She's an attractive woman as well, for forty-three. I can honestly say that she's hardworking, ambitious and not stupid either. So far, so good. But she can be quite cold.
>
> We were recently at a birthday, my aunt's eightieth. There were old work colleagues of hers there, and pensioners. Well, you know what they're like, worrying and moaning. And the woman went as far as to say,

'I've had it up to here. All this moaning and groaning is getting on my
nerves.' And then she went home. I can't understand it. (Scheuch, former
engineer at Stanex, 24 June 1991)

Karl Scheuch portrayed his sister-in-law and his brother's acquaintances
as 'career types', and he found it unpleasant, the way they ruthlessly
elbowed others out of the way to get to the top. The widespread
pejorative qualification of West Germans as 'career types' or as people
'who only live for their work' made it easier to come to terms with a
work situation in which East German employees who were already over
forty could no longer pursue a career and were afraid of losing their jobs.
    Scheuch met his brother less after the fall of the Wall than in the
preceding years, when he had had permission to visit his elderly mother
in West Berlin for two days every month. He observed, like Robert
Ruland and Horst Fröhlich, that, 'They used to say, "oh, if only you
could come over sometimes!" And now, now that we can, they don't
want to see us!' Ruland accused his relatives of speaking passionately in
favour of German reunification before the Wende, only to feel later that
the Ossis were a nuisance.

> Now the Ossis are all coming and ruining everything, ruining our lovely
> Federal Republic, which we spent so many years building up for ourselves.
> But who actually wanted the reunification? The ordinary man on the street?
> I said to my sisters, as early as '87, on my brother's fiftieth birthday 'There's
> probably more people thinking about reunification or of a united Germany
> in the GDR than over here.' Oh, and they all denied it! Well, now the truth
> is there for everyone to see. As long as the Wall was still standing they
> could say: 'Oh, if only you could come and see us sometimes, oh, if only
> they would let you travel, oh, we always have to come and see you.' And
> now we can travel, well, you ask around, how many families have fallen
> apart? Now that the poor country cousins are always coming and pestering
> them. 'Ha, here they come again! Leave us in peace!'
>   Having an opinion, and maybe even thinking that there were good
> things about the former GDR, well, that's the worst crime! If you say that
> everything's rubbish here and that in the West they're still the best, well then
> you're just sucking up to them. You have to have a bit of self-confidence, a
> bit of pride ... (Ruland, skilled worker Stanex, 18 April 1991)

Robert Ruland felt, just like Scheuch and Fröhlich, that his relatives in
the West no longer welcomed him. He explains that this rejection was
because he did not conform to the stereotype of 'the poor hillbilly', but
instead 'self-confidently' pointed out what he had liked about the former

GDR. He turned around the negative prejudice, which he suspected in his relatives and which offended him, while insinuating to them that they were suffering from 'not being the best anymore'.

The skilled worker and satirist Fröhlich, whose poems introduce some of my chapters, didn't feel obliged to paint a rosy picture of the East German way of life during discussions with his Western relatives, particularly his successful father-in-law in West Berlin. He summarised the essential common ground between East Germans as a 'jail effect'. Discussing his relationship with West Germans, he said:

> They're different. And I mean that sincerely and in a friendly way, and …
> oh, I don't know. We've had other problems for decades, and we've been
> shaped through that, and the only thing we've got left at the moment is
> an old sense of 'unity borne of necessity', in inverted commas, which
> we've lived with all these years, and it's easier to live with that than with
> well-meaning Westerners, I must admit. It's nothing personal against
> anyone. You can always make an effort but it's never the same as with the
> people – this is the jail effect I was talking about – the people who you've
> been in the same cell with and who you know how to get on with.
> (Fröhlich, skilled worker, Stanex, 16 June 1991)

In contrast to most of his fellow citizens, Fröhlich could have left the country and moved with his family to West Germany, as part of 'family consolidation' (the reuniting of families), but neither he, nor his wife, 'had the feeling that they *had* to do this.' They felt fundamentally closer to the people in the German Democratic Republic. So they were, figuratively speaking, voluntarily 'in the slammer'. For Fröhlich, the individualist, West German society was not necessarily the better alternative. So he was deeply shocked by the way in which his West Berliner father-in-law wanted to arrange a job in the West for him, even before the Wende. Fröhlich quoted him as saying, 'Well, yes, if you come, then, you know, we could just get rid of two men and put you in their place.' This offer was unacceptable to Fröhlich, but he observed that such behaviour 'has now become a reality', a reality which remained strange to him.

> We feel like we've just been released from prison and we don't know how
> it works on the outside – but the difference is that real ex-cons have
> experienced freedom before. (Fröhlich, skilled worker, Stanex, 16 June
> 1991)

Fröhlich defined himself and his acquaintances as people who couldn't know how it was on the 'outside', who had been imprisoned, and who had come to terms with the situation which had been forced upon them, something which West Germans could not comprehend. In doing this, Fröhlich did not put himself in the role of a victim. His flirting with the stereotype of the 'jailbird' was only one facet of his identity, which he used to explain his feeling of belonging to a community of East Germans.

The Stereotype of the 'Lazy Ossi'

Work in East and West constituted a topic full of anxiety for the members of the enterprises. From stories from neighbours, relatives and acquaintances, they had created an image of the job market that caused them fear. But, simultaneously, stories were also told of East Germans who were able to show their Western colleagues how productive they were, and 'that they wouldn't need to get their certificates as trained skilled workers anymore'. Tales of East Germans as an exploited workforce in the West accompanied stories about East Germans as rivals.

> They're hired for three months. Three months is often the trial period for a job. And then afterwards the next one gets taken on. Well, who's going to kick up a fuss about that? Everyone hopes that they'll be able to stay, that they're in employment again. Who's going to go to the union or the workers' council? Not the people from the East in any case, not once they've found a job with a firm in West Berlin. (Ruland, skilled worker, Stanex, 18 April 1991)

Veit Kater claimed, warningly, that many Western firms would even begin to replace the first generation of East Germans with an even cheaper workforce, to the point of employing workers from the former socialist 'brotherlands', who were willing to work for bargain prices. These stories, which could not be dismissed, and which were confirmed by others such as Karl Scheuch, who had found employment in the West shortly after the Wende, certainly contributed to the fact that many employees clung to their ailing companies in the East, especially if they were over 40 years old at the time of the fall of the Wall.

> Everyone of my age, well, you just have to look at the date of birth, and that says it all really. Fifty-two? It's just not done. Except now we're getting a situation, where, because so many young people have already left, and this surge of migration is still going on, that something's

happening that you don't even hope for or expect anymore. It is pretty unlikely that we might get a situation, when things are improving so rapidly that people begin to say, 'Well, actually, we do still need the fifty and sixty year olds …' (Ruland, skilled worker, Stanex 18 April 1991)

Walter Schuster and Karl Scheuch, who had already worked in the West, tried to define what it was that differentiated them from their Western colleagues and what, in a positive sense, defined their identity as East German workers. After the Wende, Scheuch had looked for a job in the West because Dr Schöpf was unable to identify any prospects for the Stanex firm. He had left the company, and particularly his colleagues, with a great feeling of regret, and never lost touch with them. His anecdotes of the working situation in the West were, for many of his colleagues, a guide to what they should expect in the West. His criticisms of the conditions that he came across at his new workplace in West Berlin were certainly no less harsh than they had been at Stanex.

Scheuch explained that his new boss wore out the personnel, hiring staff arbitrarily and then laying them off again soon after, in order to have the cheapest possible workforce, without taking into consideration the creation of a functioning team. Scheuch observed corporate weaknesses in dealings with customers. He described how promises were made regarding specifications that the customer had not even asked for, promises which could consequently not be kept during the manufacturing process. He characterised the owner of the firm as being hot-tempered and uncontrollable, because he vented his anger on his staff. Scheuch was one of the few new employees who were able to assert themselves within the company. He was allowed to make careful suggestions to his new boss regarding the improvement of the working atmosphere, suggestions which were, however, usually ignored.

Although they had to admit that it wasn't cold and impersonal in all places of work in the West, Scheuch and Schuster described the personal ties between the employees as being marginal. Many West Germans were, apparently, so fulfilled by their work, that their whole lives seemed to be determined by it. 'The Wessis live to work, and we work to live,' was a commonly used stereotype. They placed emphasis on the things that determined their life outside of work, and the fear of losing their jobs thus lost part of its existential dimension.

Schuster highlighted the less pronounced divide between the top and bottom of the hierarchy of employees in East German enterprises.

In the West it's completely different. In the West this hierarchy has been accepted by a lot of people. So, people who wear white collars, instead of blue ones, well, they feel like they're something better. That they're more qualified, or a better person, just better overall. And that's a real problem. Because it's just that that stops engineers and workers from cooperating creatively. In the engineering department here we've still got the advantage that we have a couple of people, who possess all the mechanical ability, and who can assemble and who have been machining for a long time. In the West there are lots of engineers, who, for example, design something, something really clever and fantastic, and then they say 'We can manufacture anything these days – so off you go, make it!' And then the strangest things get wired, milled and cylindrically ground – or whatever, I don't know – because someone takes the view that, 'It was my idea, but I don't have to actually produce it, someone else can do it.' That's an example of a typical problem in the West. (Schuster, skilled worker, Stanex, 20 February 1993)

The stereotype Schuster discusses, of the West German white-collar employee, who thinks of himself as 'something better', is based, on the one hand, on the perception that a stronger division of labour between manual workers and 'brain workers' makes the solution of practical problems more difficult, and, on the other hand, on Schuster's specific problems of status as a skilled worker. As an experienced skilled worker, who had for years worked hand in hand with the engineers at Stanex, he found it annoying that in the new system brainpower brought more prestige, and that engineers distanced themselves so much from the manufacture of the goods that they found it difficult to make useful, constructive suggestions regarding production. Although he delivered the same performance as before, his contribution was not valued as highly. As a worker, he did not have as much value in the market economy as in the planned economy, where the workers had ideologically been the supporting pillars of the workers' and peasants' state, and where their abilities for improvisation and creativity in the production process were always necessary.

Pride in one's own achievements and the desire to be accepted 'for who you are' by West German employers was also emphasised by Veit Kater, an engineer. At the end of 1991 he applied to several West German firms, with the help of a book he had bought specially, entitled *How Do I Apply for Jobs?* As he explained, he did not take any of the advice, because the code of conduct for job interviews basically suggested that one should hide one's true personality. What he wanted was to be hired for 'being himself'.

Stereotypical discourses flourished with the introduction of two different wage zones for East and West Germany. The low productivity of East German firms, and therefore the calculated low productivity of East German employees, was to be balanced by low wages, in order to maintain the firms' competitiveness. Most employees, however, were convinced that they were well-qualified strong performers. They assumed that the West Germans would have the prejudice that 'the Ossis are unqualified and unproductive', and countered with their own self-assessment, that they were efficient and disciplined.

> This idea of 'the stupid Ossis, who can't do anything, who are no use', well, you get that everywhere ... until some Western firm or organisation needs something. And then, funnily enough, all of a sudden, East Germans are very well trained people. But many businessmen, they know that people always had to find a way ... because the whole time we always had to come up with something ... to get us out of that economy of shortages ... And this defensive statement [that the Ossis can't do anything; B.M], well, I see it more as a defensive statement by long-established Western citizens, who are afraid of competition. Because we are a threat! Definitely! We really are having an effect [on wages; B.M.]. (Ruland, skilled worker, Stanex 18 April 1991)

Robert Ruland reassessed the stereotype of incompetence, which he himself had brought up, as a 'defensive statement' that was an expression of the jealousy and fear that West German workers felt towards their potential, East German, rivals on the job market. He saw that low wages in the East, along with the willingness of the East Germans to work for lower wages in West Germany than the West Germans, had created a situation of rivalry. This rivalry arose between the West Germans, who wanted to keep their wage levels, and the East Germans, who wanted to improve theirs, or who were simply looking for work.

> Well, obviously, there are people who've only been earning 6, 7, or 8 Deutschmarks up to now, and now all of a sudden they can earn 12 Deutschmarks, or13, or14. So for them this is all very lucrative. And that's for the same work that a long established citizen could legally get 18, 20 or 22 Deutschmarks an hour for, according to the official pay scale. Of course, that's a godsend for an employer, if he knows that there are those out there who will do the same for12, or14 Deutschmarks. And so there is pressure on the old citizens, a disciplinary pressure we could say, on these old citizens ... 'When you're not looking, I'll get me a couple of Ossis.' (Ruland, skilled worker, Stanex, 18 April 1991)

While Ruland constructed, on the one hand, an essentialist discourse about the Wessis who denigrated the Ossis, he was indirectly vindicating the East German workforce, who surged onto the West German job market for lower wages. Immediately afterwards, however, he abandoned this course of argument and didn't actually place the 'blame for this rivalry' on the 'little people', but rather on those (whom he signified with the vague term 'they') who profited from the abolition of the rival social and political system in the GDR.

> This contrast between the two systems has gone. The rivalry is gone; you don't need to be considerate of anyone anymore. You don't need to consider what they do in the East anymore. In my opinion, socially, everything's developing in the opposite direction. And they're calling the shots. It's this 'divide et impera', or 'divide and conquer' that brings success! Divide the little people, or even stir up hostility between them, so that you can really cement your own power somehow. (Ruland, skilled worker, Stanex, 18 April 1991)

With Ruland the construction of an East–West contrast reached such a high level of abstraction, that his arguments became increasingly unverifiable. The powers that 'stir things up between the little people' were no longer being named and so appeared to be even more threatening. Ruland was also the only one of my interviewees to speak of a colonization of East Germany. In his notebook he carried an article whose author quoted Theodor Storm, pointing out parallels between the situation of Schleswig Holstein in the nineteenth century, and that of East Germany following reunification. Ruland read aloud:

> Here: 'How history repeats itself – at the beginning of 1867, after the Prussian victory at Koeniggraetz, the two former duchies of Schleswig and Holstein were annexed by the Prussian state. Theodor Storm wrote the following lines about the system of brutal sovereignty. I found them by chance and am astonished by the way things repeat themselves. "We cannot fail to recognise that we are living solely under force. It's so much more drastic, as it comes from those whom we called for help against the Danish brutality, and who now, after they helped us deal with them, are treating us like a conquered tribe. They are throwing away our most important institutions, without asking us, and imposing others on us at their discretion."' (Ruland, skilled worker, Stanex, 18 April 1991)

Ruland found that the similarities were 'Obvious! The bureaucracy, the snobbishness, the haughtiness – all add up to arrogance – just like today.'

Above all, one parallel seemed to impose itself – the destruction of important institutions and the imposition of new ones, without consulting the citizens, just like Storm had complained about. Ruland did not question, in this context, the exploitation of the new institutional structures by the East German director of his own enterprise. Transferring accountability onto an abstract 'colonial power' excused him from establishing that link.

It is striking that the stereotypes that the East German employees used to construe these East–West differences, and that they used to deal with these contrasts, often did not characterise West Germans directly, but had more to do with the supposed prejudices that they were convinced the West Germans had developed about them. These hypothetical stereotypes were then in turn contradicted by a schematising counter-discourse, which stressed positive East German commonalities. As stereotypes are not permanently defined and immovable, and are more often equivocal and ambivalent, they can be strategically employed and can simultaneously conceal strategies and interests (Herzfeld 1997: 67).

The Idea of Socialism

'Not everything was bad for us here.' This fairly humble claim to a past that does not have to be shamefully silenced, had an effect both externally in confrontation with the new reality of the Federal Republic, and internally in the affirmation of an East German identity. It challenged my interviewees to examine their ideals of the past, the idea of socialism, and to reevaluate socialist society and their own role within the system. This self-reflection was strongest at Stanex, where arguments about worldviews and political strategies had been intense even before the fall of the Wall. Colleagues continued their political debates outside the boundaries of the company as well, as staff members Scheuch and Schuster were already working in West Berlin. The discussion amongst the staff at Stanex regarding the idea of socialism was not only an ideological debate, but rather it concerned their entire identity. It was about, as engineer Kater put it, 'remaining ourselves' in a rapidly changing social environment.

Discussions about socialism as an ideal for society took up themes that had led, in the 1980s, to tragic conflicts in the engineering department (see Chapter 4). Once again the engineer Scheuch was a catalyst. In his new job in West Berlin he missed his colleagues at Stanex and also the open, if often conflictual, working atmosphere that had prevailed in the engineering department. With experience of working in both the West and the East, he developed an interpretation of the socialist system, which he discussed with his colleagues at Stanex.

> I don't think that the experiment of socialism has been disproved. I don't believe that it's not possible. I don't believe it. Because, the fact that it went wrong in the GDR and everywhere else, well, in my opinion there were a great many reasons for that, and in the GDR, where of course I am in a position to judge, it is, or was, surely very closely connected to the fact that professional competence was destroyed, and that objections were not allowed. Isn't that so? When I was studying engineering I learned – in politics lessons, or whatever it was called – I learned that conflict encourages progress. And I assume that that's true. If there is conflict, and it must be resolved, then you'll get the best out if it in the end. The conflict must be resolved somehow. You can solve it with force as well, but, if you solve problems reasonably, then, as we always say, progress will always be a result. (Scheuch, former engineer at Stanex, 24 June 1991)

Scheuch's criticism of the social system of the GDR is based upon dialectics. Progress, as he learned at engineering school in the 1960s, is only possible when conflicts in a higher unity of synthesis cancel each other out. Indeed, what was given as an explanation for scientific and technological advances in the1960s, just as Scheuch was at engineering school, was no longer considered valid by the start of the1970s, when Honecker announced that the ultimate goal of social development had apparently now been achieved with real existing socialism. Economic and social advancement should not, in future, develop along the lines of a permanent revolution of the powers of production, rather it should be a result of evolution, an evolution explicitly and politically determined by the Socialist Unity Party (SED) (Meuschel 1992: 222). This meant the negation of conflict within the real socialist society.

For Scheuch, but also for his ideological opponent Ruland, it was the negation of conflict that ultimately destroyed the social system of the GDR. Even Ruland referred to Marxism as a philosophy which enables a complete understanding of society, but which was not able to have any effect in the GDR due to the synchronisation of opinion. On the contrary, he was of the opinion that the capitalists had become astute enough to use the knowledge that Marx could bring them for their own egotistical interests, while those in power in the GDR trampled these insights under foot.

Well, the capitalists have made full use of this philosophy for themselves. It's all in there! About labour productivity, about the individuality of people, of each person. That not everyone can be made to follow the same path, and that opposing views simply do exist ... and that diversity of opinions can lead, to some degree, to reasonable ways of living together. But that's just been trampled all over. Just as church officials walk all over their Lord and Master on a day-to-day basis. (Ruland, skilled worker, Stanex, 18 April 1991)

Ruland's idea of socialism was shaped significantly by the philosophical views of his strict Catholic parents. For him, socialism and Christian doctrine belonged inextricably together. Socialist ideas had been, according to him, subject to many variations through the whole of human history, and they were striving for more 'equality' and 'quality of life for ordinary people'.

I'm going to be bold and argue: who was Jesus Christ then? Basically he was a socialist, if you really pull apart the stories. And what did he want? Fundamentally, a better life for the poor. He was from a poor family himself, of course, a very humble little family. (Ruland, skilled worker, Stanex, 18 April 1991)

Admittedly, for Ruland there was a clear distinction between Christian ideals and the rule of the Church, which he had rejected, just as he had refused to join the Socialist Unity Party (SED).

I knew a lot of people, who really represented these socialist thoughts with their hearts, who were not in the Party either and who, for the same reasons, also did not feel drawn towards it. So I sometimes said, perhaps there are more socialists outside of the Party than there are socialists within it. (Ruland, skilled worker, Stanex, 18 April 1991)

As a 'party-less' socialist and trade union member, Ruland, bearer of the 'Fatherland's Gold Order of Merit', did not count, in contrast to Scheuch, as a critic of the regime. In 1991, Scheuch still regarded Ruland as a coward and an informer for the regime, while Ruland himself regarded Scheuch as someone who had been converted to socialism because of the circumstances of capitalism.

He's someone who had previously been one of the harshest critics – admittedly, I must say, he only wanted that things move forwards in the GDR as much as possible. He didn't want to know about socialism or anything like that before. So it was like a red rag to a bull, he went mad in discussions about the Party, and even I can understand that a little. He said, the last time he was here, really quietly to me … because we were often politically opposed, but we always had mutual respect for each other … well, he said, 'This notion of socialism, I don't think it's dead and buried.' And for me to hear that come out of his mouth, it was somehow – oh, I don't know – it was wonderful! … It does show you how this new situation has triggered people's thought processes, and makes you regret that we haven't done something more with these ideas and from these beginnings, and that we haven't done more with the GDR. (Ruland, skilled worker, Stanex, 18 April 1991)

While Ruland saw a convert in Scheuch, he felt that he himself had been confirmed as someone who was right and who didn't need to revise his view of the world. Scheuch's colleague Kater saw it differently, and asserted that Scheuch was one of the few employees who had stood by his opinions and had not aligned them with the political regime.

There are a couple of guys like Scheuch – he has always clearly said, that he's against what's happening here, but he's also taken the standpoint from the very beginning that the West is not exactly the bee's knees. … And now he's getting on with it. And he hasn't changed his opinion. He used to be conservative, now he's a leftie, officially anyway. He was so angry before and now, suddenly, he's someone who says exactly the same things that he used to, but by doing so he's so much further to the left than many others, who, all of a sudden, can only see the Christian Democrats now. But there's not many people like him. (Kater, engineer, Stanex, 25 April 1991)

'Not having to change his viewpoint' and 'to be able to remain true to his beliefs', was important during the first period of change. Over and over again my interviewees emphasised what had not changed in their view of the world. Veit Kater admired his colleague Karl Scheuch for being able to remain consistent in his views, while Kater himself, as a party member until 1989 and, in his youth, a doctrinaire representative of the regime, had to question and change his political practice several times. He ascertained that some critics of the relations of power within real existing socialism were also critical of the Federal Republic's social order, while many followers of the old regime rapidly integrated themselves into the new one.

Veit Kater's former party comrade Walter Schuster had less admiration for Scheuch and accused him of 'not having taken the right path either', meaning that he did not, in fact, engage himself for his political convictions at the time of the upheaval. Instead, he looked for work in the West and then called upon his colleagues in the East to defend the status quo there.

> And then he said, 'Keep everything the way it is.' He's gone away, and then, speaking from experience he tells us, 'Keep everything the way it is, it's better than what I am experiencing now!' What he sees as a flaw now, that of course private ownership holds a certain brutality – well, he didn't feel that at all at the time, because it wasn't like that then, because at that time it simply didn't have that effect. Now it all has an effect. Now money has a more brutal effect than before, because there are these huge differences and because of all the flat noses [pressed up against inaccessible glittering shop windows; B.M.], and now property has a stronger effect and the lack of jobs just as much, and so there's this feeling of 'not being needed' … particularly in those aged between forty-five and fifty-five or even up to sixty, when it really hurts some people. … They have not been able to appreciate the achievements (of socialism) and also they haven't contributed towards improving the situation, but have been rather more interested in fulfilling their consumerist desires. (Schuster, skilled worker, Stanex, 20 February 1993)

An overabundance – not only of goods, but also of people – characterised the market economy, according to Schuster, and it was apparently this side of the coin that the GDR citizens had not appreciated, as they allowed the old system to collapse so that they could become part of the Federal Republic's consumer society. Like many of his colleagues, Karl Scheuch had been apparently unable to recognise the brutality of a society that is organised via the medium of money until he experienced it himself. Also Schuster had tried two years before this interview, in 1991, to make 'socialism' coincide with his fundamental ideal of 'community and social cooperation'.

> Socialism was actually a positive aim in life, and capitalism the aim which actually survived, and that's still my belief now. I can take Christian beliefs and all the other world religions, and if you thin down their substance, then the bottom line is that you're left with community, social cooperation and not fighting against one another. So then, of course, the question is always, how can I realise these goals and what is my actual aim in life? It is not my aim in life to sweat away the whole day in a factory, so that somebody else can make a rich life for himself! That can't be it! My intelligence bristles against that! Just because someone is the son of a fat wealthy fellow, he can learn more, and go to a better school … (Schuster, skilled worker, Stanex, 22 April 1991)

He contrasts the actual goal of 'socialism' with the stereotype of the 'fat rich capitalist', someone for whom he would not like to work. Neither he nor his colleagues had any concrete idea of what a socialist society should look like. They were, however, in agreement about what they did not want: a monolithic society, in which contradictions were excluded. At the same time they wished for a society of harmony and not of struggle, a social togetherness. However, they did not, yet, have any practical ideas about how they intended to realise these visions. Therefore, 'socialism' remained for most of them on the level of an ideal, an alternative religion, which, if one believed in it, was able to reconcile contradictions. When the need was there to take action, and not simply to 'leave everything how it is', as Schuster accused Scheuch of doing, considerations about socialism provided no plan of action which actually related to the society in which they lived.

Images of friends and foes were challenged in East German society by the collapse of the GDR, the privatisation of enterprises and the differentiation between 'brain workers' and manual workers, and these images had to be newly defined and validated. The 'other' had to be defined as a partner or as an opponent, before it was possible, to act with or against them. The difficulty of a new definition such as this is demonstrated by the relationship between Scheuch, an engineer, and Ruland, a skilled worker. Whereas Ruland believed he had found corroboration of his view of the world and of himself, Scheuch saw, in contrast, weaknesses in his character and cowardice. The ideal of socialism that many employees at Stanex wanted to hold on to, could not be the same one that they had earlier pursued in the other reality of real existing socialism. 'Socialism' had, now, to be thought of beyond the leading party of the working class. This demanded the development of entirely new political and social strategies, about which the staff could not yet develop any common ideas.

The staff of Stanex saw themselves as the mirror image of what they had construed as the 'other'. The stereotypical discourse that they attributed to West German prejudice about East Germans allowed them to distance themselves and formulate positively who they were. The frightening abstractness of the system of capitalism was personified and came within reach in the stereotypical figure of the Wessi. This negative figure of the Wessi did not correspond to any general description of characteristics of people living in West Germany, even when individuals, such as Scheuch's sister-in-law, were concretely designated, in a specific context, as being cold and heartless. The Wessi is not a person, but rather a capitalist archetype, which the employees with whom I spoke distanced themselves from, just as much as from the contrary stereotype of the Ossi. The Wessi had ambivalent characteristics, 'he' consciously allowed the East German economy to break down, but 'he' was no self-confident, conquering destroyer, rather 'he' was afraid of competition, afraid of the Ossis. To construe the 'other' as negative or insecure helped to get one's own insecurity under control and enabled one to deal with everyday life.

Admittedly, this individual act of coping with everyday life has not led to collective action. Explanations for economic decline and the destruction of social relationships in the enterprise were expressed in an either/or sense – either the Wessis and the 'market economy' were to blame, or it was the former rulers, who had continued to hold institutional positions of power. The construction of a 'Wessi archetype' concealed the power relations within the new social structure and prevented the employees from seeing the obvious collusion between old socialist and new capitalist structures of power, and from developing their own strategies of resistance more effectively.

Notes

1. An example: on a cycling tour with an East German friend through Brandenburg we were forced by an inconsiderate driver with a Berlin registration plate into the ditch at the side of the road. My friend shouted 'typical Ossi!', whilst I shouted, at the same time, 'typical Wessi!'. We were probably both wrong to do so, only that wasn't the point, for the use of stereotypical classification served in this case to assure both of us that we had no negative prejudices against either East or West Germans and we were able to continue our cycle ride in peace.

2. Friedman differentiates three ways in which culture enters the wider social system. The first way refers to an 'objective' description of the content of life of the population 'out there', defined through their distance from 'us'. The second form of culture brings together the elements that a population uses to identify itself – for example, language or ancestry. The third concept views culture as an organiser of total life processes, including material reproduction. This refers to a 'former' social system, which can only exist outside of the present system. Culture here is not organised in order to enjoy advantages within the system, rather it should, in contrast, enable one to leave the system (Friedman 1994: 88–9). In a similar approach Stolcke draws our attention to the political backgrounds and special relationships that cultural differences create (1995: 12). She emphasises that identity, whether ethnic, cultural, national, political or gender-related, creates a relationship, which, logically, always assumes a contrast to others (1995: 12). She calls upon us to listen to the dialogue between ideologists and their supporters and to observe the economic background which cultural fundamentalism thrives upon. (Stolcke 1995: 21).

Part III

Joining the World Economy

In the initial years after the fall of the Wall, the East Germans felt they were taken over by the dominant social system of the Federal Republic of Germany. The employees traced the decisive changes in their enterprises back to the influence of West Germany's economic power, funded by West German institutions. They perceived it as a collision with a previously competing political and economic system to which people were either ready to adapt or which they resisted. The 'market' was embodied by West German consumers and 'privatization' by the Treuhand institution. The antithesis, whether as an opponent or as a goal to strive for, seemed recognizable and was personified.

As the employees in the East Berlin subsidiary of the multinational corporation Hochinauf quickly learned, this perspective on the world was merely a fraction of the new social conditions. In contrast to the employees at Taghell and Stanex, they were transposed onto a global corporate scene through the purchase of their firm by a multinational company. The question was no longer whether they felt West or East German – these categories certainly tended to dissolve into 'Owis', in other words, East Germans who earned Western wages in the West (*Ostdeutsche, die zu Westtarifen im Westen arbeiteten*), and 'Wossis', that is, West Germans who worked in the East (*Westdeutsche, die im Osten arbeiteten*). Instead, the employees were summoned to find their new identity via the multinational company, which was defined as a 'family', the Hochinauf family, by corporate ideologues. Identity, which at the time of the GDR was supposed to emerge from socialism and the 'people of the GDR', was now expected to come out of belonging to the multinational company.

The multinational company Hochinauf is like 'a racing car with constantly new, constantly better fine tuning', an expert in the field wrote in 1996 in New York. About twenty years ago the company had begun to expand and maintain a constantly changing network of production plants in thirty countries, and run maintenance branches in 220 countries. Hochinauf appeared to be a company of the future, which effortlessly crossed national borders, fitting its products with the latest technologies, developing and producing increasingly eye-catching models for its customers. Its company strategy is in harmony with cataclysmic forecasts for the future, which predict an explosive surge in global population. Rural depopulation in most parts of the world and urban growth in the so-called newly industrializing countries are factors which permeate the official company strategy and permit positive forecasts for future turnover. An increase of 700 million people in cities throughout Southeast Asia means more high-rise flats and therefore a greater demand for lifts and escalators.

The collapse of the planned economies in Central and Eastern Europe generated hope for a rise in demand for technologically valuable products and Hochinauf rapidly expanded its range of action to encompass the ex-socialist countries in order to win ground on rivals that also operated internationally. The company is continuing to develop. In 1997 the strategy of maintaining local production plants with a variety of products was abandonned. The new president announced he would transform Hochinauf into a truly 'global company', reduce the product variety in favour of flexible standard models whose standardized components could be produced on a global scale at the cheapest production sites. Plans were made to trim the forty-five factories and nineteen engineering centres down to four or five main locations by 2002.

The mental geography of the employees at Hochinauf – or so the story went – would no longer be confined to the borders of a country, but would include the network of company branches throughout the world.

Chapter 8

Hochinauf's Mission in East Germany

The expansion of the business into East Germany was supported by Western managers' conviction that, along with its economic and commercial goals, Hochinauf had a cultural, and even civilizing, mission to fulfil in the East. Against the irrationality and arbitrariness of the planned economy, as they perceived it, the managers laid down an enlightenment idea of man, based on economic rationalism, optimism and individualism. In their opinion, capitalism had clearly won the battle between the two systems. They therefore wanted to 'help the losers' make the leap into the market economy and create the basic cultural, legal and social conditions that this would require.

The civilizing mission worked through the construction of categories of identity, which were based on management philosophy. The institutional semantics of the enterprise philosophy of the multinational company played with polysemy and ambivalence. The key categories – 'competition', 'belief in success', 'sense of family'– were both contradictory and complementary. They related not only to economic calculation, but also to feelings and beliefs. What did these categories mean for those who had created them, and who reproduced them inside and outside of the company, and what effect did they have?

It had been set in writing by management that a fundamental principle of company policy was to grant everyone within the business the same chances. This was explained in the tone of a constitution:

A fundamental principle at Hochinauf is that all qualified persons must be offered the same opportunities for work, which means without consideration of race, beliefs, skin colour, country of origin, age, gender, disability or other factors not connected to the business interests of Hochinauf. (Company policy, undated)

Factors not connected to the business interests of Hochinauf were to have no effect on the careers of employees. Conversely, this also meant that the business interests of Hochinauf, which focused on finances and balance sheets, were the objective yardstick against which the staff would be measured. Hochinauf saw itself as leading the way in a global civilization of the future, one which would be based on objective economic rationality.

At the same time the purely economic rationality of this discourse was combined with another, contradictory, discourse, which alluded to feelings and used non-economic and non-maximising argumentation. When the managers of Hochinauf spoke of integrating new businesses into the combine, they expressed it as 'integrating new family members'. The idea of the 'Hochinauf Family' appeared in advertising material and on internal information bulletins for members of staff. It alluded to relationships of solidarity between the members of the company, to the experience of community and mutual responsibility. The strength of this concept lay in its emotional power and its ambivalence – it encompassed hierarchical relationships as much as egalitarian ones. It created an emotional relationship with the enterprise and implied a collective interest. It turned the universe of economic rationality into a universe of emotional uncertainty.

This 'enterprise philosophy' is part of a trend, which originated in the USA, of applying a functionalist concept of 'culture' to the company. Values and norms that influence the working relations of businesses are 'diagnosed' and 'controlled'. What is in the 'hearts and minds' of the employees – the assumption is – can and should be managed in the interests of the organization. The aim is not only submission to the goals of the firm, but also the development of personal initiatives in the interests of the firm. The employees should measure their personal success according to the contribution they make towards the success of the company; their identity should become one with that of the firm. The enterprise philosophy showed the employees the kind of behaviour and attitudes that were desired, with the implicit – or even explicit – promise that they assured their own position in society with the success of the enterprise. Flexibility, creativity, orientation towards achievement, team spirit, being aware of one's responsibilities, and, above all, identification with the interests of the enterprise are the patterns of behaviour which are demanded.

In their bestselling book for managers, *In Search of Excellence*, Peters and Waterman explain that a strong business culture creates an emotional connection to the company, and that the internalisation of clear business values can often replace formal structures. Furthermore, this business culture gives the employee the opportunity to stand out as an individual, and to adopt the philosophy and belief system of the company as an overriding system of meaning – 'a wonderful combination' (Peters and Waterman 1982: 81). The ideal employees are those that have internalised the goals, values and 'culture' of the company and who, therefore, no longer require strict controls. Within the controversy surrounding the deliberate creation of a company culture as a means of normative control, supporters argue that the staff, as much as the firm, profit from this. Employees are motivated and encouraged to achieve, and the firm offers them 'a good life', and that means a friendly, cooperative working atmosphere and the opportunity for self-realisation (Ouchi 1981). Critics, on the other hand, see a new, subtle form of tyranny in the normative control of employees. Bendix (1974) calls it a surreptitious take-over of the personalities of the workers and an attempt to control their experiences, thoughts and feelings, which in turn determine their actions. Garsten points to the intrinsic ambivalence of these organizational values that acquire different meanings at different places (Garsten 1994: 217)

It is the ambivalence of the 'civilising mission' of Hochinauf's West German managers in East Berlin which I will examine more closely: their worldviews and projects for individuals in society. How do these fit in with ideas that the employees had developed in the planned economy?

Strategies of Expansion

Priority was given at Hochinauf to the shareholders, referred to as a 'conglomeration of many small shareholders' by the Human Resources Manager, Pfeiffer. The new maxims of the corporate group in the middle of the 1990s stood in contrast to the 'social contract' of the West German social market economy of the 1970s, when employers and workers agreed on compromises in their mutual interest. In the 1990s the enterprise was no longer responsive to its staff, but to the shareholders, who had given the company their 'trust' and who had put their capital at its disposal. Shareholder value – the profits that shareholders were able to take from the company – took priority. Pfeiffer explained:

We live in competition with other firms. Internally we supply and want Italy and France etc. to buy our products. So I can't say 'Clocks go differently here, so please accept our significantly higher Hochinauf prices.' At the same time, everyone knows that we have a [majority; B.M.] partner who is in turn accountable to the shareholders. The combine ABC is a conglomeration of many small shareholders; but admittedly not in Germany, rather in the USA. Why, I wonder, why should shareholders elsewhere understand that in Germany much less profit and interest is earned for the shareholders than in other countries? (Pfeiffer, Human Resources Manager, Hochinauf, 29 August 1996)

Pfeiffer made shareholders appear as identifiable individuals, although, for the most part, shares in Hochinauf were held by impersonal institutions such as banks, insurance firms and, increasingly, rival firms and investment companies, who tie up bundles of shares, which the individual investor then does not even know s/he owns. The financial structures of dependency and the targets for profit making, which were set by the U.S. head office for its European subsidiaries, determined the business policy. The profits that each subsidiary company yielded were the parameters that measured its standing within the corporation.

Dependent on the often fluctuating market of the construction industry and confronted with strong international rivals, Hochinauf bought up large and small national manufacturers or acquired shares in them, and integrated them into the conglomerate. The task of running the firms according to local requirements was left to the local managers. As the president of the company expressed it in a newspaper interview, Hochinauf had 'left behind the idea that it was only an exporter' and now viewed itself as an enterprise with 'a large domestic market', supported by the knowledge, experience and cultural diversity of its local managers. Admittedly, the manufacture and distribution of new lifts constituted only part of the income of the firm. For example, in 1996 the majority of the company's income in Germany was obtained through the modernisation of old lift systems and the supply of services. However, the sale of new equipment expanded and secured the network of service customers, and was thus a necessary part of the business.

Only a few foreign experts were appointed to hold together this network of companies; in 1996 they were only two hundred worldwide, out of a total of 68 thousand employees. These were, however, in strategically decisive positions. Above all, they had to ensure that the finances and accounting of the local firms were structured according to the guidelines of head office, in order to provide for a detailed, central information system, transparent for the central administration of the combine. The experts also acted as mediators between the national firms and the European and American head offices where investment decisions were made. They taught the East German firms the right way of dealing with customers and ensured that the staff joined in with the enterprise's philosophy, so that 'they endeavoured, as much as they would, if it were their own company, to realise Hochinauf's vision for the future'.

An area of tension existed between the ideal of the autonomy of the individual employee and the local production line, on the one hand, and the practice of careful supervision and evaluation according to financial criteria by head office, on the other. The tension between normative and financial control pervaded the entire multinational corporation. These tensions became particularly apparent where businesses following the logic of the planned economy were to be integrated into the corporation. As Hochinauf took over the East German enterprise, the political ideals and moral concepts of West German managers met with the striving for profit of the corporation.

To summarise briefly, the Hochinauf strategy can be characterised in the following way: take over majority shares in existing firms, observe the mode of operation of the firms, and homogenise them with the multinational company. Immediately after the fall of the Wall, VEBLift in East Berlin was courted by Western companies that also produced lifts, and that wanted to ensure stakes in the monopoly for service contracts that the people-owned enterprise held in East Germany. Already before 9 November 1989 relationships between Hochinauf and the VEBLift had been good, as Hochinauf had, produced lift systems for Interhotels and party premises (*Parteihäuser*), which were then maintained by VEBLift. Parts of their lift for heavy loads had been produced by the people-owned enterprise VEBLift. Contact between the firms was intensified in December 1990 when a rival of the Hochinauf corporation announced in a press release that they would like to cooperate with the Kombinat, of which VEBLift was a part. In January 1990 the first talks took place with the three rival Western lift firms.

At the same time, the managers of the Hochinauf group appealed to the directors of the VEB and, in March 1990, invited them on a tour of the group's most modern production sites in Europe, and also to the European head offices in Paris. Hochinauf also began to actively advise its new partner on matters of company policy. Consequently, at the beginning of 1990, the corporation convinced the VEB to bring all maintenance contracts once again under their administration.

Maintenance had been an unrewarding duty under real existing socialism and the VEB had tried as much as possible to rid itself from this obligation. In March 1990 a preliminary agreement with Hochinauf was reached in which the group guaranteed 900 jobs. This promise was the decisive factor for the newly elected heads of the workers' council to favour the multinational corporation over the other candidates. During this time, the corporation began to act as an adviser for the employees and executives and attempted, as Pfeiffer put it, 'to arbitrate in emotionally loaded conflicts and debates' among the personnel. As of 1 July 1990, VEBLift, now called Hochinauf was directly under the influence of the group, which now held 60 percent of the shares. From July 1990 the group began to send consultants into the company and to put their own company name onto the firm's delivery vehicles. The manufacturing workers and mechanics doing service work were given Hochinauf uniforms.

Admittedly, the management of the Western firm had only a vague idea of the constraints under which the staff had had to plan production and investment during the planned economy. Investment planning was based on the filled order books and the idea that the types of lifts produced in East Berlin would carry on meeting demand. In the months following currency union, however, the company experienced an unpleasant surprise, as numerous orders for the old types of lift were cancelled. Stolz, the West German investment adviser, reported:

> Well no-one had imagined that the products that were made here were not actually worth much, and that they would more or less die out with the introduction of the Deutschmark and of German unity. None had counted on that happening. They had orders from customers and didn't even think that a large number of these customers wouldn't stay loyal to their contracts with Hochinauf. (Stolz, Hochinauf, 9 February 1993)

Several times Stolz emphasised that 'none had expected' that, within a short time, work would have to be found for the staff of the East Berlin production site. Hochinauf had, during the negotiations, made the generous promise to the staff to guarantee 900 jobs. But by 1991 50 percent of employees were laid off or given early retirement and production lines were closed down. Two of the GDR standard lift models were redesigned as special editions that could be integrated into Hochinauf's range of lifts. The modernisation of old GDR facilities with new doors, intercom systems and cabin interiors was a further source of income. The remaining employees were consolidated into small teams, which now assembled both the special edition lifts and the last orders for the discontinued old models. Stolz acknowledged that the company had not kept the promises made to the staff, but said, before he left Berlin in 1993, that the company had fared better than many others, where 90–5 percent of the workers were laid off.

The takeover of VEBLift by the multinational corporation was portrayed as a smooth continuation of a company history, which had been interrupted by forty years of real existing socialism. VEBLift was the lost daughter company, which was welcomed with open arms back into the growing family of companies, and all of its employees were welcomed too. This myth of continuity was, of course, purely fictional, for the West Berlin lift firm, which had belonged before the Second World War to the East Berlin company, had been bought years earlier by the multinational enterprise with its head office in the USA. But as with any myth surrounding origins, this one gave real relationships a superior meaning and disguised them at the same time.

The Missionaries of the Market Economy

Kern, the manager responsible for the Eastern European branch of the company at the European headquarters of Hochinauf, was convinced that the structure of the business organisation itself, the attitude to work, what was expected of the staff and the technology produced, was all exemplary. His explicit goal was to introduce through the Hochinauf companies new structures of working together into Eastern Europe, 'which would prepare the people in the enterprises for life in a democracy'.

The idea that Kern associated with this was to create fair conditions for competition, which would give all individuals, businesses and countries the same competitive chances. During his time as a manager in Mexico he ensured that illiterate members of staff learnt to read and write, paid for by the company. In South Africa, before the end of Apartheid, he had the wall torn down which had separated the canteens of the white and black workers. He opposed preferential treatment of family members and fought nepotism and corruption. Kern desired understanding and tolerance amongst all peoples and nationalities, and hoped to be able to realise these ideas in the global firm Hochinauf.

After Hochinauf took over the majority of the business in East Berlin, Stolz, an experienced manager, was sent into the company as a consultant who was not tied to any particular area of the business. He worked as adviser to the branch in East Berlin from 1990 to 1993. As he put it, he had arrived in Berlin 'to secure jobs in the East'. He talked to colleagues and heads of all departments, observed the flow of production, asked the staff to provide him with task descriptions and with site diagrams, calculated productivity and profitability, planned restructuring in collaboration with local managers, and drew up investment plans with them.

Stolz had worked his whole life for Hochinauf. He had experienced turbulent periods of restructuring before and knew the manager of the European head office personally. His career began in 1957 at Hochinauf in West Berlin, where he worked as a fitter, and after this he studied engineering for three years. Following completion of his studies he applied for work at Hochinauf, as he had done his earlier training there, and was immediately hired as a design engineer. He rapidly became team leader of the design engineering section and was later responsible for all the production and investment planning. At the beginning of the 1970s he was offered the chance to take over the management of the engineering section of a large escalator factory that the company owned in West Germany. Drastic business fluctuations in the manufacture of escalators, above all in the mid-1970s, made investment planning there a thankless task and caused Stolz's problematic relationships with his superior. In 1990, as Stolz's conflict with his boss grew more acute, he took advantage of the offer to move back to Berlin and to work out concepts for the restructuring of the Hochinauf branch in East Germany. Finally, he took over management of production planning in the East Berlin firm himself and began to implement his own ideas.

> I drew up a working paper, in July 1990 it was, which more or less suggested everything that we're doing here now. (Stolz, Hochinauf, 9 February 1993)

Stolz tried, as he himself says, not to appear to be a 'know-it-all Wessi', but rather to use the experiences, gathered throughout his work for Hochinauf, in the East Berlin factory. In doing this, he stood between the East German staff, who were apprehensive of him and his extensive ill-defined authority, and the West German superiors, who wanted to hear from him how the new members of staff in East Berlin saw the future of the company. Stolz introduced open-plan offices, developed ideas about machinery layout and office allocation, all of which disturbed the habitual order at the workplace. He was in favour of self-regulating work and argued that employees should work in the interests of the whole company, and not just in the interests of their own departments.

It's sort of got something to do with philosophy: I'd look for other people's mistakes first. That's only human. But actually it's better to ask what you've done wrong yourself. Did I go to my colleagues and say, 'Listen Paule, unless you give me the documents right away, then we're not going to get on very well anymore.' Or 'Dear buyer, I decided that I need this and that, and it's still not here. So can you tell me when it will arrive?' You should never say, 'I couldn't assemble the cabin frame because no-one sent me the pulley.' To me, that's like saying, 'I starved, because no-one fed me'. (Stolz, Hochinauf, 9 February 1993)

Stolz looked for and found weaknesses in the company structures and processes, paid attention to mistakes, and offered solutions, whether they were asked for or not. His presence on the shop-floor and in the other departments was sporadic, and his observations and interventions did not necessarily correspond to his formal position in production planning and budget management. Colleagues nicknamed him the 'Grey Eminence', a title which the despised director of material economy had carried, before he was laid off for Stasi collaboration. Hardly any of the East German employees came to speak to Stolz voluntarily. Only a few members of staff agreed with the style of work he was demanding from them.

I have an excellent man working for me. He works independently and makes decisions for himself. He put it brilliantly once. I was standing behind him, and he hadn't noticed. And he says, 'yeah, our fat man sits himself down and takes a scrap of notepaper and makes a couple of notes and then leaves the rest to us. And we know where he wants to go.' For me that is something very positive. And that's exactly the opposite of how it used to be here in the East. People really used to wait until the boss had given them instructions. And as the bosses, if I may be blunt, were not always experts, well then the poor staff members would be waiting forever for instructions. (Stolz, Hochinauf, 9 February 1993)

Stolz took the view that there was no problem that could not be solved and he therefore demanded a positive approach from his colleagues and staff. He made suggestions, but gave no ready-made solutions and demanded that his staff joined in his train of thought, with the ultimate goal always before their eyes. Only staff who accepted the objectives of the business, such as increasing productivity and quality, and who were prepared to work more quickly and efficiently, were able, in Stolz's eyes, to make demands on the firm.

I always say for me the statement, 'This is not possible' or 'I cannot achieve this' does not exist. I only accept, that one might say 'okay, the goals have been specified, I'll start thinking about how we can achieve them.' And then it's legitimate if that person makes demands. (Stolz, Hochinauf, 9 February 1993)

As for the question of how he would motivate employees to change their attitude if they did not believe in his goals or their achievability, Stolz would respond with, 'Preach, preach, preach. Take the people by the hand and lead them. Act as an example and join in.' Stolz invited the staff of the management team to countless individual discussions and tried to affect their leadership style and business methods. At the same time, however, he remained, personally, at a distance and therefore isolated. His valuable advice was always tied up with the crude threat that the factory would be closed down if productivity and quality did not improve.

You sit together in a department meeting and say 'We have to improve our productivity and you, workshop, you should sort it out!' And then you get asked the same question five or ten times: 'What will happen if we don't manage it?' And if you've said ten times, 'If we don't manage it, then we close the shop' and if they still don't grasp it, well, I don't know how many times it can be said. (Stolz, Hochinauf, 9 February 1993)

Insecurity was the prevailing feeling and was consciously used by management as a disciplinary tool. The East Berlin workers in the factory felt threatened by the constant accusation of not being able to match up to the productivity expectations of the Western owners. In addition, there was the vague insecurity fuelled by rumours surrounding the restructuring of the entire group, the phasing out of German production sites and the instability of the development of the building industry, which the lift manufacturing branch depended on for orders. Stolz was aware of these fears and used them as a means of pressure:

There are many employees who still have a lot of anxiety about their jobs, because they can't see ahead to what Hochinauf will be able to do in the future. And that's substantiated by the comments that people make in private, rather than out in the open. We say, if it doesn't get any better here, if productivity doesn't improve, then it is very possible that Hochinauf cannot keep up this branch. Then that means, that in the end we won't need them anymore. (Stolz, Hochinauf, 9 February 1993)

Stolz saw himself as one who took decisions about the number of workers necessary, and not as one about whom decisions were made. He was convinced that advantages could be derived from the harsh competition for jobs within the firm and from competition for market shares and profits on the world market. He saw it as an inevitable consequence that others lost out in that process, a consequence which arises in any competition, even one which has been carried out fairly.

> For a few million people in Europe, a few Japanese, and a few Americans, everything goes well, and the rest of the world gets a bad deal. Is that really our aim? We're doing something wrong here.
>
> Well, I wouldn't want anything different, because I'm an egoist, but not a hypocritical egoist. When I say to someone, that I knowingly ruin things for him, then he has the chance to respond, to take action against me. Then he can have an honest fight, like in a boxing match – 'I want to knock you out and you've got the same chance, you can knock me out too.' I live in this society and I am always going to try and get the best from it for myself, to take advantages. That's obvious. (Stolz, Hochinauf, 9 February 1993)

In response to his own moral objections, Stolz characterised himself as an 'un-hypocritical egoist', who challenges others to 'a fair, honest fight'. Stolz's words, as crude as they sound, are also based on a fiction: the idea of equal chances that had become one of the fundamental business principles of the multinational group. As Stolz's worldview of the ubiquitous 'fair fight' could be threatening also to himself, the idea of the fight was accompanied by a firm belief in personal success. Stolz believed himself to be among those who are chosen as winners through the mechanism of 'moral Darwinism', as Bourdieu calls it. 'Through the cult of the winner, who is formed in higher maths and au saut à l'élastique, fight between each and everyone, and cynicism become the norm for all action,' (Bourdieu 1998: 116).

The Hochinauf 'Family'

When Hochinauf managers talk about the integration of new businesses into the multinational company, they express it as the 'acceptance of new members into the Hochinauf family'. The concept of 'family' creates images of lasting relationships between members and the experience of community and mutual responsibility. This concept of the Hochinauf family is part of the official discourse which is communicated to all new members and which represents the external picture of the business in advertising brochures and employee information. Staff themselves also use the concept, for example, when the chairman of the workers' council requested 'family behaviour' from management. It is the indefiniteness of the term that gives it its strength. The company's portrayal of itself as a family gives the impression that it would be in the interests of all the staff to focus their creative energy and activity in a direction that is profitable for the business as a whole.

The Hochinauf brochure for employees gives details of what kind of behaviour is expected from the staff:

> Well-informed employees, who are encouraged to use their energy and initiative, will grow into a real team. Through close cooperation problems can be turned into opportunities, and with appropriate training and the support of management, they will endeavour, as if they were independent business owners, to realise Hochinauf's vision for the future and to meet the challenges of the competitors.

The employee handbook is full of ambivalence about the entrepreneurial role that the staff should fulfil. On the one hand, they should behave like 'independent business owners', but, on the other hand, should not develop any of their own goals and strategies, but instead make the business 'visions' of Hochinauf their own, and realise them. Just as mysterious as the concept of the 'vision for the future' is the request for employees to turn 'problems into opportunities'. The means for this is teamwork, in which, each person becomes part of the team, and uses, as an individual, his energy and initiative in the superior interests of the business. A real team, as the West German adviser of the business saw it, distinguishes itself through a strong selection process that sorts out the weak team members.

During the takeover of firms from the former planned economy, cultural and psychological patterns such as thoughts of success and striving for achievement, and feelings of responsibility and creativity, are directly influenced by changes in organisational structure. To remain with the Hochinauf example, in order to become a full-fledged member of the Hochinauf family, the employee must first learn to think in the ways of Hochinauf. The central concept to be learned is success and the central orientation to adopt is that every goal is achievable. By keeping the goal before one's eyes, the means of achieving that goal can be found. In the worldwide contest for markets and profits the Hochinauf employee should feel like a winner. Stolz clarified these central points of the Hochinauf philosophy by using the example of the moon landing by the Americans:

> I always use the example of John F. Kennedy. He said, 'Dear nation, I want to land on the moon by the end of the seventies. Make that the number one priority.' He didn't say, 'That's a fantasy, we're not going to get to the moon.' Instead he said, 'I want to go there' and then his people said 'My dear John F.! Right then! It will cost so many billions, and we have to do it and we will do it, and it's okay.' Do you think they would have got there if everyone had said 'what nonsense, you'll never get there!' Then none would ever have gone to the moon. (Stolz, Hochinauf, 9 February 1993)

Someone determines the goal and then those executing it find the ways and means to realise it. Stolz does not consider the possibility that the nation might have said: 'What in God's name are we going to the moon for? Haven't we got more important concerns that we should be spending our billions on?' Those executing should believe in the goals that are presented to them. The Nietzschean idea, that the will is the driving force of all action, should be used for the success of the company. In a strange reversal of the concept of will, the requirement is to believe in the efficiency of wanting. The philosophy of will has become magical. The individuals who work for the company should believe that their will influences the course of events.

Unconditional and irrational belief in success is a necessary complement to make tolerable the second central concept that should shape the thoughts of the employees, that of competition. As the employees of Hochinauf were to be in competition with not only all other lift manufacturers in the world, but also with the staff of Hochinauf in other production locations, who potentially produced more cheaply or more efficiently, they were under immense pressure, which could only be lessened if they believed that they could win the competition. As Stolz expressed it, this rivalry should be joyfully and positively taken up, because winners 'live well in our society'. 'What does market economy mean, at the end of the day? Market economy, it's competition, being better than the others in order to have the chance to survive.' He portrayed competition as the natural law of the market, which determined relationships within the business and led to businesses closing and staff being dismissed, or, to return to the Hochinauf jargon, family members being excluded.

The obligation of believing in success could not override the employees' awareness of the possibility of being dismissed. Identification with the business was thus ambivalent. Individual employees were often in a position of identifying superficially with the business some of the time, and opposing and distancing themselves from it at other times. Normative control of the individual in the organisation was not absolute (Goffman 1961). The individual always possessed the freedom to interpret his/her situation and to find his/her own meaning for it. Kunda, who analysed the relationship between business culture and the self in a globally operating technologically advanced business, demonstrated how free space becomes systematically filled with the rhetoric of management ideology (1992: 215). Where Goffman perceived spaces of freedom for the individual in the organisation, Kunda saw a 'culture trap', which combined normative pressure with persuasion and force (1992: 224). He characterised as ambivalent the distance and identification that employees felt towards the role that was offered to them in the organisation. Business culture is only an additional instrument of control in the hands of management that supplements other means that the management possesses of ensuring the cooperation of the employees.

For human resources manager Pfeiffer, 'making moral decisions' meant abstracting oneself from individual fates and giving preference to the most cost-effective production location according to the 'objective' laws of the market. He described the closing of the production plant in East Berlin, where he had been the human resources manager for six years, with the neutral expression, 'the basic conditions are changing'. He argued that the production site there had to be closed, as it was losing a million a month in 1996, due to lack of orders and structural problems. It had to be closed before the debts accumulated and the entire German branch of the business – and therefore also other production plants – were in danger.

> For me it's a question of morality; the closure of a production location with consequences for all concerned there, well – we're not fooling ourselves – at the moment it's certainly an unpleasant situation for us too. Only I wouldn't say that in West Germany it causes less dismay than in Berlin, just because they aren't as close to our workers' council as we are. I have to say that very matter-of-factly, very business-like; 'unemployment arises here or there as an outcome, or may arise as a result, and has to be judged as being of the same quality'. (Pfeiffer, Hochinauf, 29 August 1996)

When job losses or factory closures occurred, it meant that the fiction of the business as a human community – or in this case, 'a family' – disintegrated. Only arguments about financial rationality remained, which did not fill the moral emptiness.

The Hochinauf philosophy suggests a complete worldview, ready to fill the void that the ideology of real existing socialism had left behind. In the current climate of individual disorientation and economic and social upheaval, the corporate ideology proposed a new home to the 'homeless mind' (Marcus 1992: 313). The model conveyed is that of the self-made man, which places a limitless belief in the power of individual will. As Stolz expressed it with so much conviction: having this belief is not just a possibility in the enterprise, but is actually the only acceptable frame of mind.

The corporate philosophy tries to reconcile contradictions and to communicate an image of transparency, order and cleanliness, whereas under the surface there is an arbitrariness determined by profit interests. These are crucial in determining when factories are opened and closed in an increasingly rapid cycle, and when people's fates are experimented with. The missionaries of the market economy had an idea of man abstracted from personal relationships, passions and empathies, and reduced the people working to their function as a workforce. On the other hand, it placed the individual will of the employees at the centre, in order to direct them towards the interests of the company and to exploit them.

The three categories 'competition as law of the market', 'sense of family', and 'belief in success' were not only invented to persuade others, but they were shared by those who created the corporate philosophy. These categories give the apparently rational, financial code of practice a magical aspect, for the idea of controlling human matters purely by looking at the financial aspects alone is unbearable even for those who have built up the multinational corporation on the basis of these criteria. The belief that the will to success actually brings success is reassuring in an insecure world. When Kern, a valued manager who had believed in Hochinauf and considered it to be leading the way in an open-minded moral world order, was dismissed in the summer of 1997 during a restructuring of the business, this shocked the entire European management of Hochinauf.

Chapter 9

Two Careers in the
Market Economy

Making the will to succeed into a motor for success was the strategy of the personnel resources management in a campaign to influence the behaviour of staff in 1993. This campaign was to support the transfer of the production of luxury lifts in glass and chrome from West to East Berlin, together with its workers and engineers, merging with the production line there. Hochinauf accomplished what some managers jokingly called 'the German unification on a small scale'. The young head of the manufacturing department in West Berlin, originally from East Berlin, returned there now to share responsibility with his Eastern counterpart. Management of the corporation had decided that this measure would accelerate the process of adaptation of the East Berlin staff to the model of the multinational corporation and would cut costs of lift production. In this chapter I want to examine how two managers, with different professional trajectories that shaped their hopes and expectations towards the multinational corporation, enacted the corporate philosophy. This philosophy presented the new power relationships as being for the best of everyone. Was the staff convinced by it, did they endorse it, and was this philosophy therefore functioning like a hegemonic ideology in the Gramscian sense (Gramsci 1959: 344)?

A Model for Success

In the new campaign that was to explain the Hochinauf model to the members of staff, a colourful chart that looked like a comic, was distributed to all employees in managing positions, from the foreman to the head of manufacturing. In six short catchphrases it explained how a loser becomes a winner:

From losing to winning
- The losers are always part of the problem – The winners are always part of the solution.
- The losers say: 'This is not my responsibility' – The winners ask: 'How can I help you?'
- The losers see a problem in each solution – The winners see a solution for each problem.
- The losers say: 'It is possible but much too difficult.' – The winners say: 'It is difficult, but possible.'
- The losers see others as part of the problem. – The winners see others as part of the solution.
- The losers always have an excuse. – The winners always have a plan.

These short formulas seem like recipes for success and happiness. The drawings show the winners with happy faces. They all smile brightly. Those who cooperate with others, see their colleagues in a positive light and confront difficulties with optimism, contribute to problem solving, and do not have problems themselves. No one and nothing can stand in their way as they strive to reach the aims that the firm has set.

However, this presentation also shows the menacing side of the demands made on the behaviour of the staff. The losers on the lefthand side look unhappy. Their hair is sticking up. Their heads rest on their hands. The faces express fear, insecurity and stress. The message is clear: those who do not follow, those who see problems everywhere, avoid risks and difficulties, and search for excuses for their weaknesses will be the losers. Everyone is free to imagine the consequences of such an attitude: career standstill and in the worst of cases redundancy.

At first sight the cards seem like friendly, humorous advice for the staff. In reality, however, they are geared towards the exercise of power in the enterprise. To think and act in a positive way is the norm that the firm sets for its staff. It requires an intellectual and psychic self-discipline. This code of conduct seems like an answer to what the managers regarded as the weaknesses of the planned economy. The picture of the losers corresponds in fact to ways of behaving that were of strategic importance for managers in the planned economy if they wanted to secure their position and the status of the firm. When they received the planning provisions at the beginning of the year, it was advisable to underline how problematic, unrealistic and unachievable they were. In that way they prepared the grounds for a plan revision that they would apply for in the course of the year. Also, each brigade declared in the socialist competition what special effort they had made to fulfil the plan, and that it had been the shortcomings of the other departments that had prevented them from achieving an even better result.

Production had to rely on workers to improvise, as material and staff were lacking and manufacturing tools were of an inferior quality. Their special efforts were a question of good will; therefore they could not be obliged to do them, nor be made responsible for the result. Workers expected help from others only if they were on good terms with them, and this would be done as a favour, which would have to be repaid in the future.

The behavioural model of Hochinauf presented the staff with a coherent code they should identify with and according to which they should act. It was to be accepted voluntarily much like a religious discipline in a monastery. For a competitive modern enterprise it was insufficient to impose physical discipline on its members, a form of power that Foucault described as being exercised through 'constant surveillance' (Foucault 1986: 239). The code of conduct of the multinational corporation prescribed certain attitudes in dealing with colleagues and requested self-discipline in pursuing the aims of the corporation.

I want to analyse how the two competing heads of production positioned themselves with respect to the disciplinary code of the firm. I will then illustrate the complexity of how they attempted to make their self-image, values, and norms coincide with the changes they imposed and that were imposed on them.

Adaptation and Refusal

In January 1993, in the East Berlin plant, two heads of production, who had come to know the market economy in very different ways, faced eachother as competitors. Oswald, who used to be the only production manager in the plant, was in his fifties, father of two grown-up children and for over thirty years employed in the enterprise. He faced Wolpert, who had been put in place by the Hochinauf management in West Berlin to fuse the West Berlin manufacturing of lift cabins with manufacture in East Berlin. Wolpert was thirty-four years old, divorced, and father of one daughter who had remained with her mother, when he moved in 1988 to join his girlfriend in West Berlin. Shortly before the fall of the Wall he had found employment with Hochinauf in West Berlin, where he advanced rapidly from quality inspector to head of manufacturing.

Oswald had to give up responsibility for the assembling of lift cabins to Wolpert. He remained responsible for the cutting and painting of cabin elements and the manufacturing of cabin frames and lift platforms. Two new CNC machines were added to the three existing ones and four experienced workers from West Berlin came under his responsibility. They were paid according to West German wages, while he received his inferior East German salary. Wolpert became responsible for the entire mounting section of lift cabins and for ten workers and one foreman who had previously mounted cabins under Oswald's orders, and who were paid East German wages in contrast to Wolpert's other staff from West Berlin. However, the majority of the fourteen employees, who had come along with Wolpert had been initially from East Germany and had only begun to work in West Berlin after the Wall fell. Now they all received salaries according to Western wages. Wolpert's section was conceived like a production cell, which combined construction, disposition and assembly and was supposed to maintain close contact with customers.

When I visited the two production chiefs after unification of their production lines in their new offices, two details struck me in particular. Both managers had arranged their desk and the adjacent conference tables in a way that had been typical for offices in the planned economy. Reminiscent of Lenin's office in the Kremlin, the personal desk of the chief was placed at right angles to the conference table. While he sat at his desk the chief could see all his staff while they had all their eyes directed at him. Oswald's office had always been arranged in that way whereas Wolpert did not use this order while he was head of production in West Berlin and only introduced it when he returned to the East.

The second striking detail in their offices was the place they gave to the illustrated code of conduct of Hochinauf: in Wolpert's office it was the only piece of decoration on the walls; in Oswald's office it served as a writing pad.

Oswald

Oswald had welcomed the fall of the Wall with enthusiasm. This was the day he had waited for since the Wall was built. That day, on 13 August 1961, Oswald, who had studied engineering at the Free University in West Berlin, returned of his own accord to his parents' home in East Berlin. This decision put an end to his academic career. As a 'border crosser' (*Grenzgänger*) he could not get a place at a GDR university because he was regarded as politically 'unreliable'. In the summer of 1961 he got a job with VEBLift and only in 1964 was he able to take evening classes for an engineering degree, which he obtained in 1969.

In the years after the construction of the Wall, the West seemed to move further and further away as life in the GDR required all his attention. While still studying he got a job in the construction department. After finishing his degree he was promoted to become brigadier of the department. In November 1974 he was given the difficult task of planning and building a new manufacturing hall and an administrative building, which required all the guile and intelligence of the best of socialist organizers. In 1980 his career reached its high point when he was offered the job of director of production – on condition he would join the party. When he refused to join, his career came to an end and the futures of his children were also put into question. Oswald therefore saw himself as a victim of the regime despite his professional successes.

In spring 1990, when three big Western lift companies courted VEBLift, Oswald represented the employees as president of the workers' council. He had become active in the trade union section of the enterprise in spring 1990 when the old trade union cadres withdrew and the power of the old directors was questioned by the staff. Oswald participated in negotiations with the Western buyers and favoured Hochinauf's bid because they promised to continue employing the major part of the staff.

Oswald assumed that the Western management would support moves to replace the old political leadership of the VEB. He told personnel resources director Pfeiffer about the political past of his superiors and pointed out colleagues who – he was convinced – had been working for the internal security agency the Stasi. His attempts were unsuccessful and the old director was allowed to stay.

In the autumn of 1990, Oswald left the workers' council and became head of production. His first major task consisted of laying off large parts of the staff whose interests he had previously defended as head of the workers' council. In the course of the year 1991 he became more and more critical of Western management. When I talked to him for the first time in July 1991 he was already full of criticism and resentment towards the Western management. He criticized the redundancies as a breach of the promises made to him and his colleagues. He accused the Hochinauf management of miscalculating the market position of VEBLift, a mistake that now came at the cost of the workers. But Oswald did not express this criticism openly towards his superior, with whom he developed an increasingly reserved relationship. He was very generous with criticism in an informal context, but extremely reserved if it was a matter of reporting officially to superiors. Oswald explained this restraint by the fact that he feared losing his job if he stepped into open opposition. It was the same argument he always used to reject having our conversations taped. In some conversations he even asked me not to write down some of his statements.

While the management in West Berlin put pressure on manufacturing to increase the productivity level, Oswald took a defensive position and tried to defend 'his' workers against performance demands which seemed excessive to him. In 1991 the employees of his department needed more than twice as long, than provisioned by the engineering department to mount lift cabins and platforms. Oswald himself called the requirements 'unrealistic' and speculated whether similarly high demands were made on employees in West Berlin, or whether these unattainably high production standards were imposed in order to depict the employees in East Berlin as unproductive, thus justifying the lower wages for East German workers. He thought it unfair that East Germans in general were disqualified as unproductive. He quoted examples of colleagues who had left for West Berlin, had worked in lift manufacturing there to the satisfaction of everyone and had earned twice as much.

In personnel management he disagreed in many respects with the West German manager, who wanted him to take a tough line towards his staff, while he himself allowed for exceptions and showed clemency. Oswald told me that a worker of the lift cabin assembly had repeatedly stayed away from his workplace without any excuse and had been issued a warning. When he was absent for a second, shorter period, Oswald issued him a written admonition, which the personnel director transformed into a last warning before dismissing him. When the worker stayed away for the third time Oswald wanted to lay him off but the managers, moved by the worker's tears, gave him a last chance. Oswald was angered by this gesture because it undermined his and his foremen's authority toward the workers and it went against the principles he had defended when he had represented the interests of staff in the workers' council.

Oswald also had daggers drawn towards Stolz, the head of investment management. Oswald, with the three foremen who belonged to his department, succeeded in 1991 in thwarting the repositioning of manufacturing machines that had been proposed by Stolz and his staff. An amused West German member of Stolz's group described how

Oswald had showed up at the planning meeting together with his three foremen 'like cowboys at High Noon' to attack the new arrangement. The plan had been made by a young female trainee without ever consulting Oswald. 'For his information' he only received a chequered piece of paper where the new arrangement was sketched out in bad handwriting and with lots of cross-outs. A slip of paper was attached where a few words were scribbled for 'Mister Oswald' without using any proper term of address. He also received photocopies of the plans where the new positions for the machines were marked in the same bad writing. Oswald felt planning without him was impertinent and saw the slovenliness as a lack of respect.

He felt ambivalent as head of manufacturing. On the one hand, he admitted that his biggest problem was the responsibility he had to assume for a department that stood under tremendous pressure from all sides, and, on the other, he felt slighted when part of his responsibilities were taken away. As a result, he complained in 1991 that investment planning, which had been his speciality in GDR times, now belonged to West German manager Stolz. When, near the end of 1991, a second production line for escalators was opened he expected that the West German foreman would be subordinate to him, and he was disappointed when the newcomer was put on an equal footing with him. In 1993, he learned only one week before the transfer of the lift production from West Berlin that he would have to hand over responsibility for the cabin assembly to Wolpert. He believed that cutting and colouring, the two sections for which he was solely responsible, had pressure from all sides: from the engineering department that gave him the plans for the cuts too late, and from Wolpert who demanded the components for cabin assembly. He also had to fulfil orders for the electronics division in West Berlin which were in competition with Wolpert's requirements. Oswald feared – and quite rightly so – that his department would be placed under Wolpert's leadership if he fell behind with the preparatory work.

Oswald pursued a strategy to cover himself in all directions and to make himself indisputable. This went so far that, in 1991 when he was still in charge of the cabin assembly, he put small prefabrication islands, consisting of mills and drills, in the assembly shop to make the assembly independent of the prefabrication capacities of the new department manufacturing drives. As head of manufacturing he pursued a strategy that in the planned economy would have gained him the approval of his subordinates. He tried to hide his problems from his superiors and never asked them for advice when he encountered difficulties. Instead of tackling conflicts directly, he withdrew behind his working group and tried to make the foremen his allies. In the market economy, however, this attitude was not well received by the foremen. They had the impression that Oswald unloaded his responsibility onto them and that they thereby became the targets of his superiors. To take up once again the image of the conference table in Oswald's office: the order had remained the same but its function had shifted. Oswald still assembled his staff to tell them the decisions that had been taken at superior levels and then proceeded to develop collective strategies against the pressures from above. But, above all, he did it to withdraw behind the collective

His self-image was that of a defender of the weak and helpless. He found a job for a mentally handicapped unskilled worker, who had been laid off, after the worker's equally handicapped wife had implored him to employ her husband again or the whole family would kill itself. This unskilled worker increased the unproductive hours of the department and gave it a negative image. For Oswald the toughness of the current atmosphere in the enterprise showed itself clearly when three laid-off technical designers were taking leave from their colleagues with coffee and cake on the third floor, while below in the canteen the West German head workman of cabin assembly celebrated his retirement with a sumptuous reception.

> 'Eighty years of seniority sat up there at a table. The mood was so depressed that the head of construction almost did not dare to give a present to one of the colleagues who had been made redundant. One of those dismissed, an excellent construction engineer, did not have any chance of getting a job again with Wolpert because he had a stutter. (Oswald, Hochinauf, 21 April 1993)

Oswald was deeply involved in the social structures of the former VEBLift and had also made enemies as head of the socialist enterprise trade union (BGL) and later of the Western workers' council (*Betriebsrat*). This showed itself in an anonymous letter that fell into the hands of the personnel resources manager, which denounced Oswald and nine other colleagues as informal collaborators of the internal secret service (*Staatssicherheit Stasi*) of the GDR. The personnel manager first asked me about Oswald's links to the secret service; then he interviewed Oswald himself and made him sign a declaration of non-collaboration. Oswald himself told me about the suspicions that had fallen on him. He saw himself threatened and without any means of defending himself against this character assassination.

He believed that everything could be held against him whether he resisted the accusations or not. He believed resignedly that those people who used to be able to 'scratch and crawl in the past' did this today again successfully. 'Those who have made the revolution, once again don't fit in the new times. They are once again unpopular.' In spite of the fact that for the most part Oswald saw himself as a victim of circumstance, he was also an acute observer of the strategies of his colleagues and superiors. He regarded the employees who had remained in East Berlin as collectively disadvantaged: through radical redundancies, reduced salaries and higher working times.

> The difference in income – everybody knows it – is very very high. For example the Wessis work 36 hours, the Ossis 40. The Friday afternoon shift ends for the Wessi at 5 p.m., for the Ossi at 10 p.m.. (Contribution of Oswald to a public discussion, 21 April 1993)

He was also convinced that their performance was as good, if not better than the colleagues' who had tried their luck in West Berlin. He followed with great attention each remark made by his superiors that pointed to the intention of closing the manufacturing in Berlin because it was too expensive and too inefficient. At the same time he analysed the discourse about closure and redundancy as an instrument of power. In the inaugural speech of the new production manager in spring 1993 Oswald counted that he used the words 'warning to lay-off staff' eight times and the word 'dismissal' three times. 'The main instrument of motivation is fear', was the principle he had learned in a seminar for managing personnel in the market economy. His analysis contradicted the code of the collaborator who works voluntarily and is highly motivated. Instead of feeling himself on the side of the winners, Oswald shared the fears of his colleagues of unemployment and loss of status. At the end of 1994 Oswald was asked to resign.

Wolpert

Wolpert left the planned economy in 1988 to try his luck in the market economy, not because he wanted to escape the system, but because he fell in love with a West German woman. In the GDR he had not been a member of the opposition, but rather someone who had wanted big changes in his youth and who came to an arrangement with the regime when he realized that he was unable to bring about any change.

You can believe me, when I was 16/17 years old I was a member of some
youth groups and we wanted to change everything. I don't mean of course
the FDJ (Free German Youth),1 but some autonomous groups. We wanted
to change something. I also had my Rudi Dutschke[2] period – if I may say
so – in the East. Or at least we were trying to have it … but we have never
been able to change anything. (Wolpert, Hochinauf, 15 February 1993)

When he left the GDR in 1988, he did not expect that everything would
transform so rapidly. As a child he had known the privileges his father
had enjoyed as a high-ranking official. Later during his three-week
training in a foundry he experienced inhuman working conditions he
had only once seen in the science fiction film *The Iron City*. As he
discovered the social injustices and inequalities in the GDR he was less
tempted than others to idealize the regime. In particular, he criticized the
clientelist character of the regime, where relationships counted more
than individual performance. The lack of appreciation of his
achievements and the slow working rhythm bothered him in all the jobs
he did after his studies. He spent many hours without progressing in his
work. The departments in the GDR enterprises, so he claimed, worked
against, not with one another.

When he found a job with Hochinauf after moving to West Berlin he
adapted very quickly to the requirements of the Western firm. The
demands on his performance stimulated him and challenged his ambition.
He enjoyed cooperating with his Western colleagues and appreciated above
all that they more readily responded to his problems and took over
responsibilities than his colleagues in East Berlin. 'There is less competition
and more cooperation in the West than in the East', he declared.

Among the workers in West Berlin he was known and feared for the
heavy workload that he imposed on them. In 1992, during his first
months as head of manufacturing in West Berlin, he obtained an
increase in production of one and a half times. When I visited the
manufacturing department in West Berlin in December 1992 the
foremen vehemently criticized 'Wolpert's reforms'. The acceleration of
production goes at the expense of quality, they maintained. Wolpert was
dispensing with workers in places where they were badly needed. The
workers did on average ten hours of overtime a month because the dates
of completion were set within too short a term. It annoyed the workers
particularly that Wolpert called every consultation among them
'dawdling'. The work pressure worsened the atmosphere and brought
even long-established workers in conflict with the foremen. It also led to
nonsensical decisions, such as the foreman instructing the workers to

install the roof of the cabin when it was not yet wired, although wiring was considerably more complicated once the roof was mounted. Before his department was transferred to East Berlin Wolpert explained that he did not see it as his role to put enormous pressure on the workforce but to bring a new spirit of openness and friendliness into the former socialist enterprise. Indeed he treated his subordinates in West Berlin with the same openness he experienced when he came to the West and he hoped to achieve this also in East Berlin.

> Well, I think that these thirty-seven people who will go over there on the 1st of January will bring a new wind, a fresh wind to East Berlin, in a positive sense. That starts with friendliness. Well, I think there is more openness and more friendliness here. (Wolpert, Hochinauf 15 December 1992)

When Wolpert arrived in East Berlin, he was greeted by back orders for eighteen lifts and he saw himself under obligation to exercise undiminished pressure on his new East German staff. He supervised, gave advice, and checked on everything that happened in the manufacturing shop where he spent about half of his time. He also mediated problems that his department had with others and did not shy from taking a clear stance in controversies. One of the West Berlin workers remarked admiringly: 'Then off he goes. There he runs. He runs like a wind-up toy. Then he gets down to it.' He often discovered mistakes in production and the handling of machines even before the foremen did. With his tape measure at hand he proved to the workers down to the minutest detail the accuracy of his criticisms and explained to the foremen how they should reorganize work.

> I tell him: 'You have to organise yourself and your staff more effectively.' I then give him an example: 'Two people should fix a post. One of them holds the post and the other one goes to the workbench that is ten metres away, gets a screw clamp and attaches it. During all that time the other one stands there and holds the post. Then he drills a hole and the other one still stands there and looks ... well, you know such things ... If I do a job I think first, what sort of tools do I need? I put the flexible tool cart next to the workplace. I organise work in such a way, that I can work better and more efficiently. (Wolpert, Hochinauf, 15 February 1993)

His way of behaving freed the foremen from part of their responsibility. The East Berlin foreman, Bierschenk, who had previously worked under Oswald, was particularly satisfied that he could count on Wolpert's support in conflicts of authority with the workers and that he shielded him from higher management. Thanks to Wolpert he had been paid overtime for the first time without having to ask for it.

Wolpert barely hid the contempt he felt for Oswald. The two men treated eachother outwardly with polite distance after their first open argument, when Wolpert won the foreman of cabin assembly as a member of his staff. One month after the move to East Berlin Wolpert explained that the lift manufacturing should be subordinated to a single leader to remedy the coordination problems between his and Oswald's department. He thought a more centralised model would be appropriate for the situation in East Berlin; therefore he considered abandoning the production cells that brought together manufacturing, engineering and material management.

Wolpert remarked with bitterness that he had become 'tough' since his return to East Berlin and that this return had been more difficult than his exit in 1988.

> Nothing has changed here. Here they still have the same bosses, the same offices where they work. The same stores are in front of the windows and there is perhaps still the black spot on the wall where the portrait of Honecker used to hang, I am exaggerating a bit now. Well, there is not much that has changed; the enterprise has a different name but you still hurt yourself against the same barriers that you have fought in the old days. Same old, same old. (Wolpert, Hochinauf, 15.2.1993)

Wolpert saw his work in East Berlin as a mission. This became clear when he spoke about 'developing the conscience of his staff' and making sure that the satisfaction of the customer became the 'specific, understandable aim and ideal for their work'. Like a missionary who had just been converted, he returned to the old socialist authoritarian style of leadership to convey the new message and to overcome the resistance of his East Berlin colleagues. This may also explain his return to the socialist style of furnishing his office. He criticised some members of staff and Oswald with expressions that a superior would use towards his subordinates, which sounded presumptuous coming from a colleague of equal rank:

> I take it for granted that a manager knows that behind every deadline there stands not just another number, but, a real customer. That means money, jobs and so on. (Wolpert, Hochinauf, 15 February 1993)

He tried to teach his staff to link the expectations they had as consumers to what they were ready to do for their customers. The idea of a fair exchange took on an almost universal moral dimension. On a more concrete level, he argued that because the production of lifts was not cost effective, the manufacturing department in East Berlin was in a precarious situation. The production of luxurious, highly visible lift cabins only continued because they were an advertisement for the company and its craftsmanship. Wolpert had calculated:

> We lose 20,000 to 30,000 Deutschmarks on average with every cabin we make. A small firm would be broke by now. It only makes sense because there are orders behind it for so and so many escalators and so and so many service contracts that make the money for us. We in this shop, we don't earn the money. because productivity is low, the administration is too big, the engineering too expensive and because we simply have thousands of problems. We have to solve these step by step. (Wolpert, Hochinauf, 15 February 1993)

Nevertheless he appeared optimistic. 'I believe, it is simply my job to solve problems. I can never say, this does not work.' He worked twelve hours a day to set a good example because he believed that the work ethics of superiors influenced the attitude of the staff. In reality, he met with constant resistance from his East Berlin colleagues.

> They have a tremendous amount of pride and don't want to see that it was really not so brilliant here in the old days. They always say: 'Well we have done all of this before. That's not all that brilliant.' Commitments to quality, like 'my product for my hand', we've known all of that. (Wolpert, Hochinauf, 15 February 1993)

When he presented the employees with a new management tool from the west it was answered with slogans from the planned economy. When he introduced sense of responsibility and quality consciousness as new ideals, he received as an answer: 'We already know this.'

Although Wolpert integrated smoothly into the multinational enterprise and brought the results which were expected of him, he doubted whether the system that he had left, or the one in which he was now successful, could establish a satisfactory social order. He was able to present the logic of market economy as an ideal, which one had to follow, and his discourse and his behaviour fitted the image of the winner that the enterprise propagated. He admitted, however, that he sometimes felt exploited in a very subtle way:

> The fact alone that something drives me to sit twelve hours here every day and to neglect my private life, this is a subtle kind of the exploitation

which I no longer have under control nor can I steer it. (Wolpert, Hochinauf, 15 February 1993)

In conversation with me he was initially very careful about showing his political convictions. At first he only criticised the bureaucracy of the West German social system. Fusing the concepts of 'citizen' and 'consumer' he described the system as 'not consumer friendly'. Compared to the GDR, citizens had to make an excessive effort to get the social benefits they deserved, and to avoid paying too many taxes. As our talk went on, he began to refer to convictions that he had acquired in his 'leftist' upbringing in real existing socialism.

> I still believe, probably because of my leftist upbringing at the time in socialism, that there will be and can be another social system. What this could look like, I can't tell. Where market economy and socialism somehow ... or maybe this is wishful thinking, I don't know. I think that this here can't be a solution in the long run. (Wolpert, Hochinauf, 15 February 1993)

He emphasized that he was 'not in favour' of the West German system, although he benefited materially from it. In the new political system he did not see himself capable of changing things that bothered him.

> I can go on a demonstration tomorrow, I can express my political opinions as much as I like. I can also go and burn car tyres in Kreuzberg and they won't even punish me a great deal for that. But that way, I cannot change anything much, just as I was unable to change anything in the old days. We can go on discussing until smoke comes out of our heads and tomorrow we can consider initiating a political trend or movement. But it is really very difficult to change certain things. (Wolpert, Hochinauf, 15 February 1993)

The resignation he expressed when talking about political engagement stood in striking contrast to the activism and imagination he developed inside the enterprise. Speaking about his frustrations with political engagement, he unwittingly came back to his efforts to create among his members of staff a new consciousness of the needs of the customers.

Wolpert left the East Berlin enterprise in 1996 to take on a job in the central Hochinauf administration supervising the whole of Europe. This was just a few months before the decision came to close down manufacturing in East Berlin in June 1996.

Self-discipline and *Eigensinn*

By promoting the distinction of winners and losers, the Hochinauf philosophy accelerated the selection process in the enterprise between those who were successful and desired, and those who were unsuccessful and unwanted. Discipline should penetrate daily life, actions, use of time and space, and attitudes towards others. The highest form of discipline was required: self-discipline. It was not rigid and prescriptive but nevertheless pervasive and goal-oriented.

Is it possible that persons with a long work experience can consciously change their systems of values and norms because they are influenced by an enterprise philosophy? Can discipline be internalised in that way?

Foucault describes the internalisation of an external discipline as a mechanism of power that transforms individuals into subjects (Foucault 1987: 246). The subject is subjugated through control and dependency. But it is his/her own conscience and self-interpretation that ties him/her to this identity of dependence. The effectiveness of these mechanisms of power is not complete. They encounter resistance that the individual wages to defend his/her status and his/her identity against the privileges of knowledge and the tangible manifestations of this power (Foucault 1986: 246).

Oswald could not relate to the behavioural code and felt threatened by it because it did not correspond to his previous work experience. He saw himself under pressure from external forces that were outside of his control. He thought the new model of the successful winner fitted well for people like his competitor Wolpert. He called him the successful parvenu, popular with his managers and consumed by ambition. 'Wolpert's behaviour has similarities with behaviour within the animal kingdom', he judged, 'he wants to stand in the limelight and he knows how to make a career.' Wolpert himself was conscious of the precarious economic situation of his line of production and took it upon himself to solve the problems of the enterprise. However, he was also able to distance himself from the Hochinauf model and to note self-critically that he was 'driven' to work too much and to put aside his private life. As a matter of fact, neither of them actually entirely internalised the behavioural code of Hochinauf.

The acceptance or rejection of the behavioural norms, of orienting one's behaviour towards performance and success, risk acceptance and strategic cooperation, depended on many factors. The corporate model was taken as a guideline or was ignored, depending on the social and historical context of its member, on the individual and collective strategies they pursued and which convictions they adhered to. The values and norms the employees held could also be contradictory in themselves.

Social anthropologists tried to grasp the link between the imposition of cultural models and the exercise of power in colonial societies by using the concept of acculturation. Acculturation is based on two aspects: the clash between two cultures and the dominance of one culture over the other (Wachtel 1974: 125). In the process of acculturation, not only is the culture of the conquerors imposed on the conquered, but the latter also actively assimilate elements of the culture of the conquerors, without renouncing their own particularities (Wachtel ibid.).

It makes a fundamental difference whether an individual decides to adopt the behavioural codes of a new group, or whether a member or representative of a group tries to change group behaviour on the basis of a model imposed from outside. When Wolpert went to West Berlin he had to learn the codes of his new environment in order to survive as an individual in the new social context. Oswald used elements of the socialist economic culture and protected the cohesion of the staff as a group as it was endangered through restructuring and redundancies. While in 1990 he tried to push changes in the power structure and a settling of accounts with the socialist past in the name of the staff, by 1991 he felt that the new owners had used him to advance their interests. As head of the manufacturing department he became responsible for redundancies and restructuring that contradicted his political ideals and his past involvement as president of the workers' council. The code of conduct that was to give his members of staff the feeling of being individually responsible for the success or failure of the system seemed like a crude propaganda tool now as they were confronted with redundancies and restructuring, which were decided without consulting the East Berlin management. The employee as an individual disappeared behind the interests of the firm.

The values and norms of the two leaders were the result of a complex blend of diverse influences. They changed dynamically with the experiences that the two men had. Wolpert made himself the promoter of the corporate philosophy of a multinational corporation and explained at the same time the origin of his worldview as rooted in his leftist socialist education. He criticised the GDR regime because it did not live up to its own socialist ideals and because it encouraged injustice and inequality. In the West he saw his ideals of an open, friendly and cooperative society more readily realised than in the East, but he did not feel he was able to be a critical, engaged citizen either. He was, as he very vaguely expressed it, 'not in favour'.

Oswald belonged in GDR times to those who saw themselves formed by a Western work ethic. He was proud of having resisted the political advances of the GDR regime. From the multinational firm he expected a moral renewal and the settling of political accounts with the socialist enterprise regime. When this failed to occur he began to judge the Western management as authoritarian and manipulative, just like the socialist regime had been before. He saw the organisation of Hochinauf as a regime driven by fear.

For the multinational corporation the preferences, ideals, opinions and political convictions of the employees counted for little: it was their behaviour of self-discipline and flexibility that mattered. The winner can doubt the rationality of the market economy as long as he behaves like a winner. The cultural model provides a frame. The members of staff, even if they adapt to the corporate model, can remain open to a different one that may be in contradiction to the dominant model of growth and competition. Social actors possess the possibility to breathe, think and act in a more independent way than an interpretation in terms of acculturation and hegemony would allow for. The disciplinary model of enterprise culture is also, for the winner, only a direction on a stage where economic power rules.

The motivating discourse that made the self-made man the model took effect against the background of structural constraints of German unification. The historical situation at the beginning of the 1990s was to the collective disadvantage of staff who had remained in East Berlin, in contrast to colleagues who had found work in West Berlin. They were confronted with massive redundancies and special wage regulations agreed in the German contract of unification. In 1993 they had lower salaries and longer working hours. The effort of the management to attribute to the individual the sole responsibility for his or her success threw only a thin veil over the inequalities in the enterprise. It did not succeed in presenting the relations of domination as the result of a fair selection between 'winners' and 'losers'.

Notes

1. The FDJ, *Freie Deutsche Jugend* (Free German Youth) was the offical youth group of the Sozialistische Einheitspartie (SED).
2. Rudi Dutschke was a West Berlin student leader of the 1968 movement.
3. *Eigensinn* is a term used by Alf Lüdtke (1993) to designate the fact that workers were acting in many small acts of life 'according to their own minds' in spite of mechanisms of control and surveillance. Stubbornness does not really translate it.

Chapter 10

Unification and Individualization

In February 1993 the industrial trade union IG Metall fought to keep the timetable for the alignment of wages in East and West Germany. Pfeiffer, the human resources director of Hochinauf was elected as the spokesperson of the national employers' association (*Arbeitgeberverband*). On the other side the trade union nominated Schwarz, the East Berlin chairman of the workers' council of Hochinauf. The two played a prominent part in the media. Hochinauf became a symbol of the differential treatment of East and West Germans, as employees doing the same work received different wages. Workers, managers and members of the workers' council appeared on television talk shows under the slogan 'First and second-class Berliners'. According to the media, workers who were paid the wages of the Western economic zone were called Wessis, while any others were Ossis, even if many workers and employees who now found themselves identified as Wessis, had originally come from East Berlin and had found work in the West only after the fall of the Wall.

For the employees, however, this debate was a media event rather than a real drama in the enterprise, even after the employers' association had revoked the wage agreement. The gestures of solidarity from the better-paid workers were limited to their participation in a single demonstration. Even the protests of the less advantaged East Berlin workers were restrained. Why did the workforce remain so passive despite the well-organized workers' council?

Pressure to Achieve versus
Scientifically Determinable Productivity

Issues of productivity created mutual suspicion and mistrust between West German company management and the East Berlin workforce. Under socialism an increase in productivity was a component of a given plan. For the workforce failure to achieve the scientifically defined levels of productivity was always a function of flaws in the process and in the system. In the socialist competition workers were committing themselves to eliminate these flaws in order to increase production. Time recordings served to establish production quotas which the workers agreed to and which they could reasonably attain. The foremen and brigadiers were careful that the workers did not exceed these quotas by more than 15 percent. Since price was political, it had neither relation to production costs nor consequently to productivity. As price was determined outside of the enterprise by the price commission, it could be many times higher or even considerably lower than the costs of production.

Following the takeover by Hochinauf, the East Berlin plant had no internal reference points for establishing production quotas. Manufacturing was transformed from mass production of a single type of lift, where work was shared, to making specialized unique lifts with small teams of workers. Instead of wages based on piece rates, the workers now earned an hourly wage based on the wage agreements negotiated for East Germany. The marketing department of Hochinauf had calculated that the specialized East Berlin lifts should be sold at a price which was one-quarter to one-third higher than the asking price for the Hochinauf series models produced in France. The technical director in East Berlin, who was the only one among the former directors who had dared to have himself reinstated by the workforce in early 1990, calculated from this basic price the maximum number of hours necessary to assemble the lift. Yet, in 1991 this threshold was not reached. Assembly actually required twice as much time.

Productivity turned into a political issue in the enterprise, around which controversies arose and strategies were developed. In the official statements of the West Berlin company management, the low level of productivity justified the low wages for the East Germans. Stolz, the head of production planning and of investment management, repeatedly used the phrase: 'The Ossis are all lacking purpose and motivation.' The East Berlin workforce interpreted this as: 'The Ossis are lazy, so they have to be paid less.'

Although the workers' wages were not directly affected by the calculated low-level of productivity because they earned an hourly wage, they were exposed to the constant threat from the company management that production would be discontinued if they did not work more efficiently. In reaction to this vague form of pressure, they demanded that they be given precise production quotas and clearly defined task descriptions so that they could form a clear view of their responsibilities. In contrast, Stolz argued in favour of working independently and said that employees should show an interest in the factory as a whole, and not just in their individual department.

> The dispatch department might say, for example: 'OK, we are shipping everything that is ready.' 'How much material have you received?' 'Nothing!' 'Well, you did not send anything, then.'
>
> Now we can ask the other question: 'What have you actually done so that the material finds its way to you in time for dispatching?' There are also people who say: 'That's not my job. The other guy has to bring it to me.' (Stolz, Hochinauf, 9 February 1993)

Whispers in the corridors went: 'Here we are cheaper than Turks.' – 'Turks wouldn't work here for that money.' The workers as well as the head of production, Oswald, surmised that particularly unfavourable contracts and prep work were unloaded from West Berlin to East Berlin. For example, for producing platforms as prep work for luxury lifts in West Berlin, the technical department allocated 1,126 working minutes at Eastern wages in summer 1991; the workers, however, needed at least 2,600 working minutes even under ideal conditions. The workers now asked themselves whether the workers in West Berlin had in fact managed to finish the products in this short period – or in even less time because they were paid according to the Western wages. They questioned whether time had been allotted in West Berlin for producing the entire lift, and whether the most unprofitable and time-consuming aspects were passed on to East Berlin. Rumoured sightings of contract wage slips from West Berlin, where the old times had been crossed out and replaced with new times, which were harder to achieve, added to their suspicions.

At the workers' council meeting in March 1991, the workers' council chairman, Schwarz, also restated these worries. He asked the head of human resources:

> Dr Pfeiffer, in this respect we would like to have an answer to the following question: the shortening of the production times for Hochinauf

internal contracts, which up to now have been produced at Hochinauf in West Berlin, is explained to us as a function of price. How is this possible, as, given the same production times, the price would be lower because of our lower wages? Whenever the workers' council tried to claim its right to participate in setting the production times, it received a veiled threat that if we do not manufacture the goods at this price, the contract will be passed on to other firms.

He proceeded to add, in keeping with the Hochinauf philosophy:

Is that behaviour appropriate in the Hochinauf 'family'? (Workers' council report of the enterprise meeting on 13 March 1991)

By using the term 'Hochinauf family', the chairman of the East Berlin workers' council was taking full advantage of a phrase belonging to the standard emotional vocabulary used by the Hochinauf corporation: as a member of this family he had the right to judge the management by its own moral standards, particularly when economic calculations were obviously made at the expense of the East Berlin plant. At the same time, the reference made by the workers' council to the management's 'family sentiment' contained a certain measure of the mockery with which this term was received in the internal discourse of the East Berlin employees.

Neither the head of production, Oswald, nor the employees working in the assembly of lift cabins believed that the jobs they had been assigned could be completed in the allotted time. Stolz's calls for increased productivity and achievement were also met with scepticism from the workforce:

Whenever I said anything, although they listened wide-eyed, inside they were telling themselves: 'Right, he's just lying to us again.' Especially on the subject of productivity. When I said: 'Gentlemen, we are going to have to complete that in half the time', they would listen to me, but not believe a single word of it. (Stolz, Hochinauf, 9 February 1993)

Since they had no benchmarks for comparison and had to rely on pure speculation to gauge the equivalent performances of their colleagues in West Berlin, in spring 1991 the employees took the initiative of recording the production time. From their experience in the planned economy, they assumed that there would be objective and scientifically proven standards of productivity and they therefore wanted to develop fixed criteria for productivity. They wanted to establish that in the case where production quotas were not met, the 'blame' did not lie with manufacturing, but instead could be traced to a lack of supplies, difficulties in the coordination between departments, mistakes made by the engineering department, and dated machinery for cutting. They hoped 'realistic standards' would be drawn up and pressure on the production plant thus alleviated. In spite of having no experience of recording times under REFA conditions (REFA: Association for work structure, industrial organization and corporate development), two employees of the East Berlin process planning, along with a workers' council representative (who had taken a REFA course for this specific purpose), took time recordings in the various areas of the plant in summer 1991 using a wristwatch to count the seconds. The West Berlin company management allowed these recordings to be made, even though Stolz critically noted that the process planners should concern themselves with optimizing the production process rather than taking time recordings.

The West Berlin workers' council was uneasy with their East Berlin colleagues' efforts to establish objective criteria for productivity and hence to reduce the improbabilities of the market economy. It was clear to the Western workers' councils that no time standards were to be allowed for wages paid by the hour and no time recordings at all should be made. Permitting time recordings for an hourly wage meant, for them, that management was being handed another means to apply pressure and that they were indirectly accepting time standards which were not included in the wage agreement and as such were beyond the control of the workers' council.

The West Berlin workers' council was in favour of a piecework wage because this option gave the committee more weight in discussions about performance quotas and working conditions. They no longer had these rights with an hourly wage, and random amounts of pressure could be placed on the individual worker. The workers' council had formed the idea that for a piecerate worker, who manufactured small series or single pieces, the margin of freedom under verifiable working conditions was larger than that of a worker on an hourly wage, who had a foreman standing over him, continuously pushing him to work more.

Of course, both the West and the East companies want to introduce hourly wages as far as possible. We, the workers' council, no longer have then any right of participation in matters concerning performance standards. This only exists for piece rate work or bonuses. In other words, I can in principle put more pressure on a worker to work harder if he is earning an hourly wage than I can on a piecerate worker – and we know that from experience.

And he won't get one single weary mark extra for it, even if he works considerably faster. And a piecerate worker who has a performance quota of a hundred minutes, he can finish the job in a hundred minutes, he can finish it in seventy, he can finish it in sixty. And if he stands around smoking a cigarette and a foreman shouts at him, he only has to say: 'What's your problem? I'm going to finish it all on time. I've got one hundred minutes to do it. I don't have to finish it in forty!' But a worker on an hourly wage doesn't have this chance at all because he doesn't have anything to compare it with, no benchmark or anything. The foreman is always barking at him: 'Quicker, quicker, quicker!' (Workers' council, Hochinauf West Berlin, 17 February 1993)

By taking time recordings East Berlin workers wanted to establish the benchmarks that they lacked in order to be able to estimate their own value as workers in the job market. Time recording caused the workers to begin to define their relationship with employees from other departments in financial terms, for example the relationship with the engineering department which delivered incorrect plans. The workers employed in the assembly of lift cabins enjoyed seeing mistakes made by the engineering department when they were discovered by quality control. They remarked: 'We'll write that all down and send it in an invoice to the engineering department. They'll have to pay us for it!' They went on to ask themselves how the company managed to survive financially at all when such mistakes arose time and time again so that they would be working on a single lift for over three weeks. The company could not simply raise the price.

The workers began to consider the connection between production time and production costs. They simultaneously began to judge according to financial criteria the relationships to other departments and to their colleagues. In the search for concrete criteria by which they could estimate their own value, they discovered some of the essence of market economy: the abstraction from the worker as a person and the assessment of the worker according to the financial net worth resulting from the work. Their attempt to assure themselves of their status in the factory and of their value as people was bound to fail because 'structural uncertainty' was a function of the 'flexible' management of Hochinauf. Uncertainty was part of the market economy, a system they had to come to terms with.

'Paradise' or the Successful Integration

The enthusiasm of the East Berlin workers who had found work at the Hochinauf plant in West Berlin stood in striking contrast to what workers at Taghell, Stanex and Hochinauf in East Berlin felt. They all complained that the market economy had encouraged the competitive struggle among employees for the sparse number of jobs available, had delivered unlimited power into the hands of the managers and had destroyed the margins of freedom at work

The workers in West Berlin told me that in comparison to their old jobs in the GDR, the working conditions at Hochinauf were like working in 'paradise'. A worker who used to work at a cellulose factory in Dresden even described it as 'heaven on earth'. Any anxieties they felt before beginning work in the Western factory, which had been shaped by socialist propaganda, proved unfounded in their eyes. Their new colleagues in West Berlin had not set upon them, excluded them or discriminated against them, but had accepted them:

> We always thought: 'Everyone there just looks out for themselves and tries to outdo the others.' That was the ideology that we had been fed. That wasn't at all true. Just the opposite. I went into the assembly hall in the first week. Everyone asked: 'So, how's it going? What are you up to?' They also explained how it all worked to me. They were enjoying it. You could have a chat there. You could go and get a coffee. Everything was free and easy. (Clausen, skilled worker, Hochinauf, 11 February 1993)

He and his colleagues explained unanimously that there was less competition among workers in the West. What they felt had been the competitive nature of the GDR enterprise arose from the attempts of each department to become autonomous within the enterprise. Departments avoided taking responsibility for mistakes in economic management and in order to resist the many demands made on them (Merkens et al. 1990). Workers felt the target-oriented cooperation between the Hochinauf departments to be motivating and pleasant.

Yet they also had considerable difficulties in familiarizing themselves with a style of work that left the workers completely responsible for the organisation of their work. Altmann, who used to work at VEBLift in East Berlin, described the difficulties he faced as well as the satisfaction he had:

> The first year was very difficult for me because I wasn't at all used to being responsible for everything myself. I didn't particularly shine in the

workplace in those days. First of all, they tried to teach you how to work independently. In the old days you would have had a little helper, in other words, a brigadier who arranged the order of the parts that came to you, put your papers together and also always told you plainly: 'Now you're doing this.' Now you just get a number from the old man. 'Have a look at 8606', for example, and you go to collect your parts, get a plan for them and then get started building the thing. Anything else that comes up in the meantime, you deal with it yourself. That's a great way to work because first of all time passes quickly. You aren't doing the same stupid thing all the time, but if everything fits together well, you are pleased about it. You can get something done. (Altmann, skilled worker, Hochinauf, 11 February 1993)

Before Altmann started assembling specialized lift cabins at Hochinauf, he used to work in mass production at VEBLift, organized with Taylorist methods. Work was so dull and easy that workers could screw the parts together, even 'with their heads drowned in drink', as another worker put it. At Hochinauf he got to know 'the end of the division of labour' (Kern and Schumann 1990): restoring not only know-how to production, but also independent work. In assembly in West Berlin, the worker alone was responsible for collecting the plans and parts that were necessary. If any mistakes were made, it was their responsibility to find a solution. In such cases, they turned directly to their colleagues in the cutting department for minor corrections, or they used the machine themselves without consulting the foreman.

In the economy of scarcity, workers faced with insufficient materials and spare parts had developed improvisation skills, which were rewarded even if the end result was mediocre. These skills were required in the new work organisation, but the results had to be greatly improved. The skilled workers were intentionally given leeway to develop their own approaches to solving problems and in addition they were responsible for the solutions. This meant that where engineering specifications were less detailed workers proportionately assumed more responsibility. They received help from the old head workman Herrmann, who was attributed the informal, but widely acknowledged, role of interpreting the construction engineers' sketches and often of correcting them too. The workers turned to him if they had difficulties. It was his practical experience they ultimately relied upon.

By accepting this new style of working as part of the novelty they were confronted with in the West, the workers in West Berlin were in contrast with their colleagues in East Berlin. Their counterparts in the East defended themselves against the fact that technical drawings and quotas were no longer as detailed as they were at the time of the planned economy. They did not want responsibility for jobs that had previously been done by technologists, quality assurance inspectors, brigadiers and construction engineers. Because in the past, construction drawings showed even the type of welding joint to be performed, now they felt uneasy when receiving only rough sketches that obliged them to make their own decisions on how to go about doing the job.

The most significant difference, however, that Altmann's colleague Bogner experienced with his work in the VEB, was what he defined as the 'social dimension'. He felt that he was now taken seriously as a person by his foreman, whom he described as a 'person he respected' and as 'totally human'. He especially appreciated that Reimann was not an unforgiving person, but relatively fair and even prepared to admit to his own mistakes. He organized social activities for his workers, dealt with their problems and acted as a mediator.

> It always depends on how you compare things. About technical things he has no idea at all. But he is in some way or other the foreman and has authority. When I saw him for the first time with his half-glasses, how he looked me up and down: 'Well, you have done things that way until now … ok, but now come with me.' With me always following behind. But in spite of all that – he is human, a real person.
>
> When you're dealing with a real person, you always get more out of it. He doesn't get on your nerves with some sort of meaningless nonsense. He comes along and says: 'So, listen, you're going to do that, that and that. You'll have to get it done today. That's it. We have to have it ready.' That's how it is. And then you go and do it and he leaves you in peace and everything. It is still really obvious with the East German foremen how they go on and on about some meaningless nonsense because they are so used to doing it before. (Bogner, skilled worker, Hochinauf, 11 February 1993)

The workers appreciated the small gestures of consideration that Reimann showed and the time he devoted to them: bringing them coffee when they did overtime in the afternoon, buying them curried sausage when they stayed late into the evening to finish a particularly important task. It was Reimann's idea to store the leftover pieces of stainless steel plate and sell them, to fund their social evenings. In Altmann's opinion, the good atmosphere in the department boosted his colleagues' motivation and resulted in them being willing to do overtime.

> The social dimension is in place and that's the most important part. The environment has to be right and the motivation arises from that. Then

you'll be ready to do not only eight hours, but also ten hours, nine hours, ten hours. Then you do some things that aren't allowed, for example, well. ... Shit! We've all done it some time or other! Then you've not only been at work for ten hours, but you've done twelve or fourteen hours. If the workers' council had got wind of that it would have been enough grounds for dismissal. But you worried about the fact there was a deadline and the customer was due for final inspection. (Altmann, skilled worker, Hochinauf, 11 February 1993)

Altmann's motivation to work appeared to fit perfectly into the corporate philosophy. The good relationship he had with his colleagues and superiors motivated him to put special effort into his job and to make the aim of the firm for meeting customer requirements his own. Although he was aware that he showed a lack of solidarity doing overtime which the workers' council had not sanctioned, he spoke about his illegal overtime as if he had pulled one over on them. He seemed to expect the workers' council to act like the enterprise trade union committees of the GDR, which would look after his personal concerns, such as favourable wage agreements, working conditions, holidays and parking spaces. He also viewed the committee as a superior authority representing interests other than his own. Hence, he did not feel obliged to keep to the negotiated regulations regarding overtime, which were intended to distribute work more fairly in society.

The enthusiasm the workers from East Berlin demonstrated was not necessarily shared by their Western colleagues, who were considerably more reserved in their comments about work and the social life at the plant. Furthermore, the electrical head workman, who was devoted to his career at the factory, stayed away from all social evenings with colleagues because, as he said, he could not bear certain colleagues once they were drunk. As if to illustrate this point, his colleague Warnher boasted about the sheer amount of alcohol he could consume and claimed he was lazy and stupid and only worked because the foreman put pressure on him to do so. The stories they told conformed to the critical and informed discourse of insiders, which Herzfeld refers by with the term 'cultural intimacy' (Herzfeld 1997). Both workers portrayed themselves in a way that emphasized their individuality, or their 'stubbornness' (*Eigensinn*, as Lüdtke 1989) calls it. One of them did this by showing his colleagues that he was not one of their drinking mates, and the other by showing that he neither conformed to the requirements set by company management for creative responsible work, nor did he want to. The uncritical zeal of East Berlin workers who had only come out one to three years ago might indicate that they did not yet share this cultural intimacy with their Western colleagues and therefore were not able to develop a counterimage. That is to say, beyond the official corporate ideology, they could not perceive those aspects of identity that were painstakingly concealed from the outside world, because they were contradictory or showed weaknesses. However, it is the knowledge of these contradictions and weaknesses that provides the insiders with their certainty of being part of a community, and with their familiarity with the power mechanisms in place (Herzfeld 1997: 3). The only superior whom the newcomers openly criticized was Wolpert, the head of production, who had in fact come from East Berlin. The tone and informal ways of addressing him that Altmann adopted when voicing his criticism in front of all the other workers shocked his West German colleagues, who advised him to behave more carefully with his superior.

Collectivity and Competition

After tough negotiations, the West Berlin workers' council secured advantageous conditions for the workforce, that was transferred from West to East Berlin: much higher income, fewer working hours and better holiday pay than their East German colleagues. They were able to secure their right to a Western wage with bonuses while their Eastern colleagues only earned the Eastern wage, which at the end of 1992 was in real terms only half that of the transferees.

The Western board of directors initially intended to set up production with West Berlin workers under separate management and in parallel with the production employing East Berlin workers. Along with changes in the composition of the board in December 1992 came a whole new concept for the production plant. The new director of production decided to discard his predecessor's concept and put the two production lines side by side despite the differences in wages. This decision was reached a week before the production was relocated.

Since VEBLift was taken over, investments had been made worth 15 million DM. Among the improvements were a new CNC break press, a CNC punch and a CNC-controlled shear. When the two production lines were joined, the amount of machinery was doubled. Now there were two CNC punches and two presses, which were to be put to full use. In his concept for restructuring manufacturing, the head of production planning and investment management, Stolz, had planned to have the shelves with materials placed directly between the workplaces in order to shorten the distance from each workplace. The new director of production criticized this idea and had the workshop changed again. The unified manufacturing department was under a considerable burden because of too much machinery and because production had fallen behind schedule to such a degree that eighteen lifts were still waiting to be assembled.

In January 1993, two days after the West Berlin lift assembly was transferred to East Berlin, the East Berlin workers' council was in for a surprise. Fifteen new workers from West Berlin entered the office and announced they would refuse to work if the new gas heating system in the manufacturing hall was not checked and repaired immediately. Smoke had been coming out of the heating system for weeks and emitted a strong smell of gas. Although the old-established East Berlin colleagues had already complained about the bad smell and about headaches too, they had neither laid down any concrete demands nor announced an ultimatum. The East Berlin workers' council commented:

> Nobody here would have hit upon the idea of refusing to work. Perhaps as an afterthought, but then nobody would have thought about saying: 'Come on. Let's go to the workers' council and tell them that we're not doing any more work until there is some sensible heating system here.' (Glaser, workers' council, Hochinauf, East Berlin, 10 February 1993)

The workers' council interpreted the lack of initiative from the old-established East Berlin workers as habit:

> We're more or less used only to keeping our heads down the whole time. That's because we tell ourselves, well, it doesn't make any sense, especially if it's coming from the East ... You don't want to say too much here or you get your fingers burnt. (Glaser, workers' council, Hochinauf, East Berlin, 10 February 1993)

The East Berlin workers did not turn to the workers' council very often with their matters of concern. Conversely, the elected members of the workers' council only appeared on rare occasions on the shop floor, as the West Berlin workers noticed straight away. They criticized the workers' council for not introducing itself to the newcomers:

> I must say something, this workers' council never even passed through the factory once. In West Berlin they were different. At least someone or other would come. One would come by at least every two days and you could have a word with him. I've been here for one and a half weeks – and

up to now nothing whatsoever. Not even to introduce themselves.
(Altman, skilled worker, Hochinauf, 11 February 1993)

The new arrangement, starting from spring 1993, created conflicting roles for the workers' council. Once the lift cabin production departments had been brought together, the East Berlin workers quickly reached the same level of productivity as their Western counterparts, without being on the same wage, however. The dissatisfaction of the East Berlin workforce intensified. At the same time in Berlin Brandenburg regional wage negotiations took place. The chairman of the East Berlin workers' council was appointed the regional representative of the metalworkers' Union IG Metall. In opposition the human resources director, Pfeiffer, represented the employers. As a result of the failed negotiations the employers' association (*Arbeitgeberverband*) revoked the wage agreement in February 1993.

At the first meeting after the revocation of the wage agreement, the workers' council demanded that Hochinauf set its own rules, bringing the wages and salaries of the East Berlin employees up to the level of the West. This motion was rejected by the executive board. While the West Berlin IG Metall representative was inviting the employees to join a demonstration which was taking place a couple of streets away, the human resources director forbade the employees to leave the factory floor. After some hesitation the West Berlin employees joined the colleagues from East Berlin and deserted the factory floor. Only a few weeks later, however, on the first walkout the workers on Eastern wages were alone. Their better-paid colleagues continued to work and even the members of the workers' council did not show their faces until five minutes before the buses set off for the demonstration. Most of the East Berlin workers left the factory quietly as they tried not to be noticed. They reported their leave to the East Berlin foreman, who was on Eastern wages himself.

Wolpert, their new superior, described the difference between the employees he brought along from West Berlin and those in East Berlin in terms of their relation to authority. In his opinion, the West Berlin workers felt more confident in contradicting their superiors because they could assume that their criticism would be accepted as a creative idea.

It occurred to me quite clearly when I arrived in the West that there isn't as much deference to superiors here. That is also due to this democratic upbringing and how people grew up. That calls of course for a totally different type of respect for superiors. I think that in the East, I'm talking here at the absolute extreme, there is much more of this servile

submissiveness towards superiors than in the West. Maybe calling them
'servile' is going a bit far, but I'm pointing in that direction. (Wolpert,
Hochinauf, 15 February 1993)

He suggested that the workers had this 'old habit of keeping your head
down and getting on with things'. In keeping with socialist practices
their harsh criticism of superiors was expressed secretly. In private
conversation they called the old director, who had led negotiations with
the Western buyers, a puppet and compared the influence of the West
German manager, Stolz, with that of a former director of production,
who had been dismissed because of his cooperation with the Stasi. They
questioned the quotas and orders they received from the West Berlin
management and did not accept them as readily as their colleagues who
had been in West Berlin. They considered them arbitrary and
frightening because in the demands for more productivity there was the
explicit threat of dismissal or factory closure. Fear was part and parcel of
Western management style, or so the East Berlin workers and their East
Berlin superiors unanimously believed.

The old-established East Berliners confronted their new Western
superiors with caution; and also Wolpert was in this category since he
had come with the production from West Berlin. Jokes and funny
remarks went to and from the East German foreman Biermann and the
old-established fitters of the lift cabin assembly, while any discussion
with the West German foreman was solemn and to the point. In direct
confrontations with their East Berlin superiors the workers did not shy
away from giving their opinion, but they would not voice an opinion in
public meetings. Holding back their full working potential, as well as
participating in strikes and demonstrations, was understood, if not in
fact supported, by the East Berlin superiors who were also paid Eastern
wages. The sudden rise in productivity, reaching the same level as their
West Berlin colleagues, can be regarded as a form of self-affirmation
against the feeling of being second class. East Berlin employees working
side by side with better-paid colleagues from West Berlin wanted to
show them what they were capable of.

The workers from West Berlin felt their independent style of work
restrained by the old structures ingrained in the East Berlin factory. They
returned, as they put it, 'back into the primordial soup' of the structures
of the planned economy, where the departments worked against instead
of with each other. A constant irritant was Spohr, the foreman of the
cutting department, who in the workers' eyes was foreman Reimann's
opposite. During the reorganization of production, he had to clear away

his panopticon, but his ways of dealing with people antagonised the workers from West Berlin. Altmann, who knew him from the times of the planned economy, called him a 'slave driver'. He described his style of leadership as arbitrary, violent and suspicious. Spohr did not appreciate it when workers from assembly came into his department searching for parts for their lift cabins and insisted on his department doing the prep work for them quickly. The assembly worker Bogner described a typical argument:

> The metal sheets for the foundation were not ready. They hadn't been done right and Hermann had to approach Spohr again to tell him that we still really needed these sheets. We needed them by four in the afternoon at the absolute latest. I went over and said: 'Mr Spohr! We need the parts now. Are you ready with them yet? Can I get them?' He gives me a right stupid look, he's a bit cross-eyed too so I wasn't really sure where he was looking. He looks at me and says: 'We're not a bakery here, you know.' (Bogner, skilled worker, Hochinauf, 11 February 1993)

The tension rose during 1993 between the employees who now worked side by side for different wages while doing the same amount of work. The East Berlin workers set boundaries to their area of competence, closing it off from that of their new colleagues, and refused to offer them any assistance. The workers from the West Berlin plant felt they were being scrutinized by their colleagues and hardly dared to have a break. When difficulties arose they would turn only to their colleagues from West Berlin. The level of earnings became a secret which divided the employees into Ossis and Wessis and which thwarted any attempts at solidarity. Altmann, who before the Wall fell had worked at VEBLift himself, now took the data protection law as an argument:

> When they pestered me I said: 'Look, my foreman doesn't even know how much I earn, and you don't have to know it either if my foreman doesn't. Because, you know, there is this data protection law … That stuff with comparing wages has long gone and now we've got this data protection law. Sorry that I'm being so awkward but that's the way it is now. I could not say anything else to that.' If I told them what I'm earning they'll fall flat on their faces. (Altmann, skilled worker, Hochinauf, 11 February 1993)

In order to justify his self-centred behaviour, Altmann turned the value system on its head: he characterized the work relations in the West Berlin factory as 'open and personal' and then withdrew behind the impersonal 'data protection law'. Although he feared the envy of his colleagues who earned a lower wage, he refused to regard this envious attitude as legitimate since 'equalizing' was long gone and the new social system was no longer founded upon equality. He explained the lack of solidarity in his behaviour by saying that the others had had their chance to try their luck in the West. It was their own fault they remained behind in the East.

> All that whining isn't going to achieve anything, but I always tell myself again and again: 'What you did straight after the Wende, anyone else could have done exactly the same'. (Altmann, skilled worker, Hochinauf, 11 February 1993)

The discourse of the workers was full of inconsistencies. On the one hand, they demonstrated they had now understood that egoism and the pursuit of egoistic interests was the key to success in the new social order. This is how they accounted for setting themselves apart from their East Berlin colleagues. On the other hand, they expressed their disappointment that the small and the weak were now excluded from the advantages of the new social order.

> We used to dream about freedom, about how much better life could be, somehow more relaxed, simpler. Whether everything is really so much simpler now, that's another thing altogether. Well, we noticed that to live well in this society, you need money, otherwise everything begins very quickly to teeter on the edge. If you have enough money, you are an honourable person. If you are always honest, pleasant and nice, and don't have a lot of money or you are unemployed or something similar, then you are the poorest soul. That's what we learned very fast. (Bogner, skilled worker, Hochinauf, 11 February 1993)

Bogner did not use this insight to take collective action in the interests of his less privileged colleagues, but he, like his other colleagues from West Berlin, first and foremost avoided endangering his own position.

While the observable behaviour of the East and West Berlin workers on the shop floor seemed to converge, their motivations differed. The East Berlin workers compared the relationships they had had with their superiors in the times of the GDR and said, that back then they did not need to feel frightened and nor did they show any particular reverence for their bosses. Now, they repeatedly emphasized, 'fear' was the motivation that kept them working. They felt the new organisation of work had been 'forced onto them' and they had not internalized the new discipline. They respected it out of fear of losing their jobs.

In contrast, the workers from the West Berlin factory stressed that they did not fear their superiors. Like their East Berlin colleagues, they complained about the greater pressure of work, and escaped by having a break behind half-finished lift cabins or disappeared for longer periods looking for materials or tools. They were willing to work overtime even if this exceeded the amount stipulated in law. The workers who had found work in West Berlin regarded the way work was organized as correct and 'natural' for the market economy. They therefore also accepted the necessary work discipline and internalized it. They were also less frightened of making mistakes.

Whereas the employees in East Berlin had no criteria for comparison and therefore could not accurately assess the pressure exerted by the executive board, the East Berliners who started in West Berlin entered a functioning system with established criteria. Their interests were also represented effectively via the workers' council. East German workers, whether they were now working in West or in East Berlin, showed a rather more distant and passive behaviour towards the workers' councils. In East Berlin they hardly ever approached the workers' council if they had a problem. In West Berlin they took for granted the material advantages that the workers' council gained for them.

The Hochinauf workforce was ultimately incapable of combined action because they interpreted the company reality in different ways. The executive board had managed to direct the energies of the workforce towards drawing comparisons with one another. The East Berlin workers' council had made an effort to find solutions for the East Berlin factory in cooperation with the factory management, but was pushed by the West Berlin workers' council into a role of more intensive confrontation. As a result, the initiatives of the East Berlin workers' council did not coincide with the priorities of the employees. The oldestablished East Berlin employees wanted above all to secure their jobs, but nevertheless were envious of the privileges of the newcomers from West Berlin. The workers from the West Berlin factory did not want to be envied but still wished to safeguard their acquired rights that gave them advantages over their colleagues. Consequently, the colleagues focused their aggression on each other, instead of fighting together for the continued existence of the plant.

The experiences of both groups of employees since the fall of the Wall exemplify the difference between the efforts required to adapt and orient in a foreign country and those required when your political and economic order collapses. When workers and employees decided to live and work in West Berlin they entered a strange but functioning economic and social system. When the order of the planned economy collapsed, the East Berlin workforce was confronted with a strange new order that was defined from outside. The former decided to enter a foreign country as foreigners. The latter became – like many GDR citizens – foreigners in their own country.

Conclusion

In this book I wanted to explore the concepts of society with which working people in East Berlin enterprises embraced the fall of the Berlin Wall. I have shown how ideas about changing the structures and mechanisms of real existing socialism emerged beyond the official ideology and what effect they could have in the transformation of East German society. My second purpose was to analyse the power structures in the enterprise and place them in relation to the mechanisms of ideological control in the planned and market economies.

The following points sum up how ideological control worked in GDR enterprises: in the planned economy plan fulfilment served to secure political dominance. Political conformity of the population had priority over economic results. In socialist competition the official ideology had to be formally reproduced. Individual opinions were undesired and not asked for. However, testimonies to ideological submission were not supposed to be simply copied or repeated, but actively created. The authors of wall newspapers and brigade diaries were meant to write their own texts masqueraded as presenting their own opinion, without having the possibility of expressing criticism.

The people I met in the East Berlin enterprises were, however, mostly self-reflecting, critical individuals, who contemplated right and wrong in society and their own part in it. While they reproduced the official ideology to the outside world, they also had an alternative version to the official visions of the world. They tried to assure themselves of their *Eigensinn*, their own ways of thinking and to reveal the perversion of those in power by exposing the failings of the infallible, the faith of the atheists and the ignorance of the know-it-alls. Yet the reluctant performance of socialist uniformity was efficient. Even if they were not convinced by the official ideology, their ritual subordination demonstrated that they did not see any realistic alternative to the prevailing dominance (Scott 1990: 66). The small individual escapes did not change social reality. The hidden critical discourses needed an audience in order to become a means of resistance (ibid.: 118).

While spontaneous utterances of discontent, ironic comments and political jokes had been an inherent component of socialist life in the enterprise, political debates, which had hardly existed previously, became more open and explicit from 1988. Some claimed from the official social organisation their right to a different opinion and withdrew because they were no longer prepared to devote themselves to a political order that did not trust them if they dared to think for themselves. They wanted to make themselves heard and acquire their own image of the world beyond the walls of the GDR. The slogan: 'We are the people', which inspired the Monday demonstrations (*Montagsdemonstrationen*) in Leipzig in autumn 1989, emphasizes the essence of their plea. Instead of believing in absolute truths laid down for them by the party, the citizens now wanted to participate themselves in the decisions which shaped their lives. The hidden discourses had gone public.

In this phase of rapid social transformation, the experience gained by the employees from the routine of socialist everyday life scarcely enabled them to judge the far-reaching consequences that their decisions would have in the market economy. Since the parameters of their actions changed constantly, they developed strategies, which were meant only for the short term, but which could nonetheless bring about consequences with a much wider impact.

The moment of action, even during the most radical social upheavals, is also always a moment when the current contexts of social life are replicated (Giddens 1987: 76). This perspective is already contained in the much-cited observation by Marx: 'People make their own history, but they do not make it just as they please; they do not make it under circumstances chosen by themselves, but under circumstances directly encountered, given, and transmitted from the past' (Marx 1969). In their actions, people can rely on experiences that they may have had in the past, and are limited by the conditions that they have created themselves through their actions. This does not mean that they cannot also react innovatively and creatively to new situations. Even patterns of behaviour appearing to be 'socialist heritage' or 'culture' are frequently in fact direct responses to new social situations (Burawoy and Verdery 1999: 1–2). Whenever people in East German enterprises used language and symbols from the socialist period, it did not necessarily mean that they were clinging to the past, but they were applying concepts with which they were already familiar in their daily actions to new purposes and endowing them with new meaning.

Although the ambitions and preconceptions of the employees in the enterprise were initially developed from actual socialist practice, these ideas also reached beyond this practice. They linked their own experiences to notions of social justice and order, which referred to a wide spectrum of social theories that they directly or indirectly came into contact with. Among these theories were those conveyed in schools of Marxism-Leninism about the role of production workers in the GDR, or the wisdom attributed to the American Plains Indians, which were popular in GDR youth groups, or the teachings of the Catholic Church. These theories do not necessarily provide a framework within which people think (Sabel 1982: 18), but out of various, often even contradictory theories, a 'little tradition' (Tambiah 1970) develops which responds to concrete life situations. The spectrum of these theories expanded markedly with the opening up of East German society.

The preconceptions of the market economy that circulated in East German enterprises before the fall of the Wall had very little to do with the world beyond the GDR. Most workers believed their enterprise could survive in the Western marketplace if the deficiencies of the planned economy were put right, productivity increased, and product quality improved. Even as late as 1990, a scarcity of goods and high demand were taken for granted. A saturated market was inconceivable before the Wende, and at first no rational explanation could be found for the sudden disappearance of product demand. This lack of orientation, expressed in the frequently heard sentence: 'We need someone to come in and change it all round', arose out of this discrepancy between conception and reality. At the same time, the employees criticized the fact that the initiative had been taken away from them and that the new instruments of control were now back in the hands of the old rulers. They had the feeling, that time was flying by much too fast for them to be able 'to build upon what had been achieved'.

Members of the enterprises Taghell, Stanex and Hochinauf were present in autumn and winter 1989–90 as active individuals or as groups who voiced differences in opinion in the enterprise. They primarily targeted representatives of socialist power, the enterprise directors, party secretaries and trade union committees of the enterprise. Nevertheless, their conceptions of democracy and participation quickly lost touch with reality when the institutional restructuring of East German society headed towards the West German model and democratization effectively stopped at the factory gates. The trade unions did not represent East German workers effectively in the privatization process, especially in the decisive year 1990. If they did not resign immediately from their posts under pressure from the workforce in the first weeks after the fall of the Wall, the socialist directors were officially granted managerial positions by the Treuhand. The 'comrade' (*Genosse*) directors of the planned economy became managers or even owned enterprises in the market economy, often helped along by good personal connections with employees in the Treuhandanstalt.

At this point the link between property and power becomes crucial. Property is institutionalized dispensation of power, which following the Wende was handed over through the state-owned trust company Treuhandanstalt from the 'people' to various private institutions as well as private individuals. In the first few months following the fall of the Wall, the employees and also the directors considered the people-owned enterprises they worked in as something they owned. When the property rights were transferred to private individuals, they conferred numerous instruments of power, which the employees in East German enterprises soon recognized to be to their disadvantage. The instrument for acquiring power over others was now no longer one's position in the political hierarchy, but ownership of property or an alliance with the owners.

In East Germany only a handful of enterprises were taken over by the employees and an even smaller number of these actually continue to exist today. Although the Treuhandanstalt was receptive whenever employees presented concepts for financing the takeover of their enterprise, most of them did not even try. To all but a few of the employees, ownership of a company in an entirely unfathomable economic situation hardly seemed worth fighting for – yet many of the enterprise directors saw the struggle as thoroughly worthwhile. The old rulers infused the new institutions of the market economy with their personal affairs: the director of Stanex by creating a political niche for himself in the company, the director of Taghell by going on wild capitalist ventures. Any power in the hands of the workers' committees, which were still weak and inexperienced in representing the interests of the workforce, was rapidly taken away. Workers' representatives were promoted to become heads of personnel resource management or, alternatively, to head of production and then had to deal with laying off the same staff they had previously defended. The employees, who had sought a fundamental transformation in East German society and had enthusiastically welcomed the market economy, were unable to redress the influence of those who held institutional power at the time of the GDR and who resisted any attempt to weaken their status.

Institutional innovation in the enterprise, the transfer of people-owned enterprise assets into private ownership, production for the free market, separating politics from economics, the removal of the social role of the enterprise, the introduction of accounting procedures that assessed the activities in the company according to profit criteria – these were the consequences of the Wende. They affected the workforce in completely different ways depending on their level of education, on their status within the enterprise, their age, their political convictions, their position in the hierarchy and on the particular moment when the assessment was made.

Those who saw advantages in an institutional renewal, expected performance-related pay, technological and organisational conditions for high-quality work, and solutions to blockages caused both by bad planning and by irregularities in production flow (*Stürmen und Warten*). They wanted rational planning, the end of political control and paternalism, of reproducing socialist ideology in socialist competition and the deposition of the socialist leaders and of the party organs in the enterprise. They were convinced that they would be able to show their true abilities, once the inflated administration by elite party members had been removed from power.

In the years following unification, however, there was a rise in the number of workers who were sceptical about the consequences of institutional change, some even to the extent of rejecting it. They regretted the loss of job security and were frightened of finding themselves unemployed or of the entire company going bankrupt. The well-qualified production workers rejected the more pronounced differentiation between technicians and production workers. Tighter control at the workplace, particularly in time management, replaced the socialist pact for plan fulfilment between management and production workers and was felt as an encroachment upon personal freedom. As a result of the highs and lows of incoming orders, phases of heavy production pressure, followed by quiet periods of short-time work, still continued to alternate. This frustrated those who had believed that such fluctuations in the workflow had been overcome with the end of the planned economy. Competition and exclusion, justified by the ideology of losers and winners, contradicted the conceptions many employees had of fairness. They felt it was unfair that 'socialist leaders' could become owners or managing directors and that in such positions they had the means to rule over the enterprise in a more authoritarian manner than ever before. Some employees were surprised about the loss they felt with the disappearance of their political and ideological significance as

Werktätige (the working people) who have an important contribution to make in society. Many who feared for their jobs had the impression of being superfluous in the new society.

Along with losing their job security, the workers also lost a certain freedom to do what they wanted in the enterprise. In the planned economy, they were mainly interested in expanding their margin of freedom in the enterprise, while in the market economy they were predominantly occupied with securing their jobs. The extent to which discipline had penetrated the enterprises in the GDR was not as deep as in the Federal Republic of Germany. The employees had not internalized the production discipline to the same degree – this was also because many did not regard the system of the planned economy as a rational one, and hence repeatedly tried to escape from it. The forms of surveillance and control in the GDR enterprise were more authoritarian and more hierarchical than in the Federal Republic, but they were less efficient too. At Stanex and Taghell the directors could use their new positions of power as company managers to be even more authoritarian than before.

The modern Western management ideology, as it was disseminated at Hochinauf, aimed at having the employees identify themselves with the company's interests and commercial goals and internalize the necessary work discipline. In modern Western factories, over the last thirty years, workers have been given more responsibility for the way they structure their own work. They were to be convinced and not forced to achieve a better performance at work (Burawoy 1979: 182, Kern and Schumann 1990). The ideology of harmony in the company masks, however, both the divergence of interests within the company and the stubborn individuality of the workers. Normalization – the process where a strict work discipline becomes part of everyday life in the company – was not the result of the management's and workforce's discovery of a common interest in keeping the company going. It was the outcome of a long tradition of surveillance and punishment, which in modern industrial societies has become more and more subtle and created an unspecified feeling of fear among those under surveillance (Foucault 1986: 241). Fear did not arise through direct repression, but through an atmosphere of general uncertainty about jobs and the future of the enterprise. At the same time, the model of the winner implied that the key to success was the belief in success. This model created the illusion that the individual in the company would master his/her own fate as long as he/she relentlessly pursued the interests of the company and motivated colleagues to do the same.

Just like the official ideology during the time of the planned economy, this homogenous model offered by the multinational corporation was accepted, changed or possibly rejected by its employees. Yet, it did not function in the same way. Whereas the reproduction of real socialist ideology in socialist competition was primarily calculated to elicit lip service from the employees to their political conformity, the company philosophy was intended to influence and control actions and decisions that were economically significant. Money, success and power played a part in gaining acceptance of the 'winner's' model of identity in the market economy. Fear for one's livelihood and insecurity complemented their effect.

The corporate philosophy of the multinational firm was oriented most of all towards increasing productivity in subsidiary companies in Eastern Europe. Employees in competition with all other manufacturers both within and outside of the company were meant to see themselves as winners who could achieve the highest productivity and the best results. The arguments and strategies, though, which the multinational enterprises used to try and maximize productivity while minimizing labour costs, contradicted the socialist idea of productivity as an objective, scientifically determinable category that took into account both the performance of the worker and the technical equipment at their disposal. Productivity had been a key word in the planned economy too. Hence, the imperative for ever-increasing production within an ever-diminishing time span was nothing new for the workers and the employees. In the planned economy, however, raising the level of productivity had a political-moral dimension of being a collective contribution of the workers to socialist society. Yet, in the multinational enterprise 'productivity' was a benchmark whereby the 'value' of the individual to the enterprise could be measured. On the one hand, the company philosophy aimed at the abstraction from personal relationships, passions and identities and reduced those who worked in the enterprises to nothing more than their role as part of the workforce. On the other hand, it placed the individual will of the employees, which was supposed to be directed towards – and be instrumental to – the interests of the enterprise, at the centre.

The company philosophy affected all the groups, albeit in different ways. It acted as a threat of exclusion. It acted as a model for the mental code of conduct and as a means of legitimizing a system of competition where everyone was pitted against everyone else. The message was: 'Whoever does not believe in being the winner, has only themselves to blame.' Whoever failed had not believed firmly enough, was not good enough and therefore could only blame him/herself for being rejected.

Faced with the individualistic model of the market economy, which emphasized competition and responsibility, the employees tried to apply a 'We model' that provided them with authentication. While the social structures were undergoing radical changes, they were attempting to remain consistent in their own beliefs. In contrast to the collapse of national socialist society, which had to be enforced from the outside and where the Germans were the ones who had to change, the central problem throughout the process of German unification was 'remaining who you are' in a society that was rapidly changing. In this regard, the workers I talked to construed West Germans as 'others', who were radically wrong in their judgement of East Germans. This construction of a capitalistic archetype of Wessi allowed for many inconsistencies. In the fictional dialogue with the 'other', the fact could be underlined that the Ossis were not lazy and incompetent, as a fictive Western counterpart would have insinuated, but represented serious competitors, and besides this they did not strive for competition like the Wessis, but for 'a sense of community and fraternity'.

Socialism in the GDR had failed, and along with it a tightly connected model of society, which claimed to represent an absolute truth. With its emphasis on the mechanisms of selection in the market economy, the new social system of the Federal Republic likewise concentrated on a simple truth, which, for the sake of economic rationality, 'colonized' the network of communicative relationships in a more penetrating and crippling manner (Habermas, quoted in Gorz 1990: 153). However, we have not reached the end of the story. Even if the dominant economic rationality manages to penetrate as far as the everyday world of those who are excluded from a life in employment, new open or hidden alternative discourses develop. During excursions and drinking in the pub, the colleagues and former colleagues, like at Stanex, continue to debate – in an endless search for a society that not only permits different opinions, but also gives those opinions that are distinct the chance to make a difference.

Bibliography

Abu Lughod, L. (1993) *Writing Women's. Worlds*. Berkeley: University of California Press.

Aderhold, Jens and Joachim Brüß et al. (1994) *Von der Betriebs- zur Zweckgemeinschaft. Ostdeutsche Arbeits- und Managementkulturen im Transformationsprozeß*. Berlin: ed. Sigma.

Anderson, Benedict (1983) *Imagined Communities: Reflections on the Origin and Spread of Nationalism*. London: Verso.

Arendt, Hannah (1986) 'Communicative Power'. In: Steven Lukes, (ed.) *Power*. Oxford: Blackwell.

——— (1995) *Macht und Gewalt*. München: Piper.

Arndt, Katrin (1997) *Die Einheit der Wirtschafts- und Sozialpolitik im Interesse der Arbeiterklasse und aller Werktätigen – Grenzverschiebungen in einem ostberliner Betrieb*. Unpublished master thesis at the Institute for Ethnology of the Free University of Berlin.

Asad, Talal (1993) *Genealogies of Religion: Disciplines and Reasons of Power in Islam and Christianity*. Baltimore: Johns Hopkins Press.

Bahro, Rudolf (1977) *Die Alternative. Zur Kritik des realexistierenden Sozialismus*. Köln: Europäische Verlagsanstalt.

Barth, Frederic (1978) 'Scale and Network in Urban Western Society'. In: Frederic Barth (ed.) *Scale and Social Organisation*. Oslo: Universitetsforlaget.

Belwe, Katharina (1979) *Mitwirkung im Industriebetrieb der DDR*. Opladen: Westdeutscher Verlag.

Bendix, Reinhard (1974) *Work and Authority in Industry*. Berkeley: California Paperbacks ed.

Berking, Helmuth and Sighard Neckel (1991) 'Außenseiter als Politiker'. *Soziale Welt* 42(3).

Berliner, J.S. (1988) *Soviet Industry from Stalin to Gorbachov*. Aldershot: Edward Elgar.

Bloch, Marc (1994) 'Für eine vergleichende Geschichtsbetrachtung der europäischen Gesellschaften'. In: Matthias Middell and Steffen Sammler, *Alles Gewordene hat Geschichte*. Leipzig: Reclam Verlag.

Bornemann, John (1992) *Belonging in the Two Berlins: Kin, State, Nation*. Cambridge: Cambridge University Press.

———— (1997) *Settling Accounts*. Princeton: Princeton University Press.

Bourdieu, Pierre (1994) *Die feinen Unterschiede*. Frankfurt: Suhrkamp.

———— (1998) *Contre-feux*. Paris: Liber-Raisons d'agir.

Burawoy, Michael (1979) *Manufacturing Consent*. Chicago: Chicago University Press.

———— (1985) *The Politics of Production: Factory Regimes under Capitalism and Socialism*. London: Verso.

———— (1996) 'Industrial Involution: The Russian Road to Capitalism'. In: Birgit Müller (ed.) *A la recherche des certitudes perdues … Anthropologie du travail et des affaires dans une Europe en mutation*. Berlin: Cahiers du Centre Marc Bloch.

Burawoy, Michael and Janos Lukacs (1992) *The Radiant Past: Ideology and Reality in Hungary's Road to Capitalism*. Chicago: Chicago University Press.

Burawoy, Michael and Katherine Verdery (eds.) (1999) *Uncertain Transition. Ethnographies of Change in the Postsocialist World*. Boston: Rowman.

Bust-Bartels, Axel (1980) *Herrschaft und Widerstand in den DDR Betrieben: Leistungsentlohnung, Arbeitsbedingungen, innerbetriebliche Konflikte und technologische Entwicklung*. Frankfurt: Campus.

Castel, Robert (1995) *Les métamorphoses de la question sociale*. Paris: Fayard.

Clegg, Stewart (1975) *Power, Rule and Domination*. London: Routledge and Kegan Paul.

———— (1979) *The Theory of Power and Organization*. London: Routledge and Kegan Paul.

Cohen, Anthony P. (1994) *Self Consciousness. An Alternative Anthropology of Identity*. London: Routledge.

Connerton, Paul (1989) *How Societies Remember*. Cambridge: Cambridge University Press.

Darnton, Robert (1991) *Der letzte Tanz auf der Mauer*. München: Carl Hauser Verlag.

Dilley, Roy (1992) 'Contesting Markets. A General Introduction to Market Ideology, Imagery and Discourse'. In: Roy Dilley (ed.) *Contesting Markets. Analysis of Ideology, Discourse and Practice*. Edinburgh: Edinburgh University Press.

Dittrich, Eckhard (1992) 'Bürgergesellschaft und Probleme der ökonomischen Transformation'. In: Martin Heidenreich (ed.) *Krisen, Kader, Kombinate*. Bonn: Sigma.

Edeling, Thomas (1992) 'Zwischen bürokratischer Organisation und Gemeinschaftskultur: Der Januskopf des DDR-Betriebes'. In: *Soziologen-Tag Leipzig* (1991). Berlin: Akademie Verlag.

Elster, Jon, Claus Offe and Ulrich K. Preuss (1998) *Institutional Design in Post-Communist Societies. Rebuilding the Ship at Sea*. Cambridge: Cambridge University Press.

Engler, Wolfgang (1995) *Die ungewollte Moderne. Ost-West Passagen*, Frankfurt: Suhrkamp.

Etzioni, Amitai (1990) *The Moral Dimension. Toward a New Economics*. New York: The Free Press.

Foucault, Michel (1975) *Surveiller et punir. Naissance de la prison*. Paris: Maspero.

——— (1977) *Sexualität und Wahrheit*, 1. Band, *Der Wille zum Wissen*. Frankfurt: Suhrkamp.

——— (1986) 'Disciplinary Power and Subjection'. In: Steven Lukes (ed.) *Power*. Oxford: Blackwell.

——— (1987) 'Wie wird Macht ausgeübt?'. In: H.L. Dreyfus and Paul Rabinow, *Jenseits von Strukturalismus und Hermeneutik*. Frankfurt: Athenäum.

——— (1987) 'Warum ich Macht untersuche'. In: H.L. Dreyfus and Paul Rabinow, *Jenseits von Strukturalismus und Hermeneutik*. Frankfurt: Athenäum.

Friedman, Jonathan (1994) *Cultural Identity and Global Process*. London: Sage.

Garsten, Christina (1994) *Apple Worlds. Core and Periphery in a Transnational Organizational Culture*. Stockholm: Stockholm Studies in Social Anthropology.

Geertz, Clifford (1975) 'Thick Description: Toward an Interpretative Theory of Culture'. In: Clifford Geertz, *The Interpretation of Cultures*. London: Hutchinson.

Giddens, Anthony (1987) *La constitution de la société*. Paris: PUF.

Goffman, Erving (1961) *Asylums*. Garden City New York: Anchor.

——— (1959) *The Presentation of Self in Everyday Life*. Harmondsworth: Penguin Books.

Golczewski, Mechthild (1981) *Der Balkan in deutschen und österreichischen Reise- und Erlebnisberichten: 1912–18*. Wiesbaden: Steiner.

Gorz, André (1990) *Kritik der ökonomischen Vernunft*. Berlin: Rotbuch Verlag.

Grabher, Gernot (1992) 'Instant-Capitalism. Western Investment in Eastern German Regions'. Paper prepared for the meeting of the sub-group Transnational Corporations and the European Peripherie of the RURE program, Kopenhagen, 3–6 September.

Gramsci, Antonio (1959) *Oeuvres Choisies*. Paris: Éditions sociales.

——— (1975) *Quaderni del Carcere* (4 volumes). Torino: Guilio Einaudi.

Grünert, Holle (1997) *Beschäftigungssystem und Arbeitsmarkt in der DDR*. Opladen: Leske und Budrich.

Habermas, Jürgen (1990) *Die nachholende Revolution*. Frankfurt: Suhrkamp.

Hann, Chris M. (1993) 'Social Anthropology of Socialism'. In: Chris Hann (ed.) *Socialism*. London: Routledge

——— (1998) 'Introduction: the embeddedness of property'. In: Chris M. Hann (ed.) *Property Relations. Renewing the Anthropological Tradition*. Cambridge: Cambridge University Press.

Hart, Keith (1982) 'On Commoditization'. In: Esther Goody (ed.) *From Craft to Industry*. Cambridge: Cambridge University Press.

Hart, Keith. 1999. *The Memory Bank*. Money in an Unequal World. London: Profile Books.

Hayek, Friedrich A. von (1988) *The Fatal Conceit. The Errors of Socialism*. Chicago: The University of Chicago Press.

Hein, Christoph (1989) *Der Tangospieler*. Berlin: Aufbau Verlag.

Herzfeld, Michael (1992) *The Social Production of Indifference: The Symbolic Roots of Western Bureaucracy*. Oxford, Berg.

———— (1997) *Cultural Intimacy. Social Poetics in the Nation-State*. London: Routledge.

Hilbig, Wolfgang (1992) *Aufbrüche. Erzählungen*, Frankfurt am Main: Fischer Taschenbuch Verlag.

Himber, Günter (1980) *Westliches Management aus DDR-Sicht*. Erlangen: IGW – Deutsche Gesellschaft für zeitgeschichtliche Fragen.

Hirschhausen, Christian von (1994) 'Du combinat à l'entreprise. Une analyse de la nature du combinat socialiste et des restructurations industrielles post-socialistes en Europe de l'Est'. Doctoral thesis presented to l'École Nationale Supérieure des Mines de Paris, 10 November.

Hirschman, Albert O. (1970) *Exit, Voice and Loyality*. Cambridge, MA: Harvard University Press.

———— (1984) *Engagement und Enttäuschung: Über das Schwanken der Bürger zwischen Privatwohl und Gemeinwohl*. Frankfurt: Suhrkamp.

———— (1989) *Entwicklung, Markt und Moral*. München: Hauser.

———— (1992) 'Abwanderung, Widerspruch und das Schicksal der Deutschen Demokratischen Republik'. In: *Leviathan* 20(3).

Hofmann, Michael (1995) 'Die Leipziger Metallarbeiter. Etappen sozialistischer Erfahrungsgeschichte'. In: Michael Vester, Michael Hofmann and Irene Zierke (eds.) *Soziale Milieus in Ostdeutschland*. Köln: Bund-Verlag.

Honecker, Erich (1984) *Arbeitermacht zum Wohle des Volkes*. Berlin: Dietz Verlag.

Huinink, Johannes and Mayer Karl Ulrich (1995) *Kollektiv und Eigensinn. Lebensverläufe in der DDR und danach*. Berlin: Akademie Verlag.

Jacquard, Albert (1995) *J'accuse l'économie triomphante*. Paris: Calmann-Lévy.

Joas, Hans and Martin Kohli (1993) 'Der Zusammenbruch der DDR; Fragen und Thesen'. In: Hans Joas and Martin Kohli, *Der Zusammenbruch der DDR. Soziologische Analysen*. Frankfurt: Suhrkamp.

Joost, Angela (1993) 'Über die Bedeutung von Selbst- und Fremdbildern für die innerdeutscher Verständigung'. In: Karl Otto Hondrich *Arbeitgeber West, Arbeitnehmer Ost*. Berlin: Aufbau Taschenbuch.

Kern, Horst and Michael Schumann (1990) *Das Ende der Arbeitsteilung?: Rationalisierung in der industriellen Produktion: Bestandsaufnahme*. München: C.H. Beck.

Kornai, Janos (1992) *The Socialist System. The Political Economy of Communism*. Princeton: Princeton University Press.

——— (1995) *Highway and Byways. Studies on Reform and Post-Communist Transition.* Cambridge MA: The MIT Press.

Kunda, Gideon (1992) *Engineering Culture.* Philadelphia: Temple University Press.

Lambrecht, Christine (1989) *Und dann nach Thüringen absetzen.* München: DTV.

Lenski, Gerhard (1986) 'Power and Privilege'. In: Steven Lukes (ed.) *Power.* Oxford: Blackwell.

Lepenies, Wolf (1992) *Folgen einer unerhörten Begebenheit. Die Deutschen nach der Vereinigung.* Berlin: Siedler.

Lötsch, Manfred (1989) 'Sozialstruktur der DDR – Kontinuität und Wandel'. In: Heiner Timmermann (ed.) *Sozialstruktur und sozialer Wandel in der DDR.* Saarbrücken: Dadder Verlag.

Lozac'h, Valerie (1999) 'Local Elite and Power Relations. A Comparison of Two Towns in East Germany'. In: Birgit Müller (ed.) *Power and Institutional Change in Post-Communist Eastern Europe,* Canterbury: CSA.

Lüdtke, Alf (1989) '"Die große Masse ist teilnahmslos, nimmt alles hin …" Herrschaftserfahrungen, Arbeiter-,Eigen-Sinn' und Individualität vor und nach 1933'. In: Busch and Krovoza (eds.) *Subjektivität und Geschichte. Perspektiven politischer Psychologie.* Frankfurt:Nexus.

——— (1993) '"Ehre der Arbeit". Industriearbeiter und Macht der Symbole. Zur Reichweitesymbolischer Orientierungen im Nationalsozialismus'. In: Alf Lüdtke (ed.) *Eigen-Sinn. Fabrikalltag, Arbeitererfahrungen und Politik vom Kaiserreich bis in den Faschismus.* Hamburg.

Lukes, Steven (1990) 'Marxism and Morality: Reflections on the Revolutions of 1989'. *Ethics and International Affairs* 4.

Maier, Harry (1993) 'Die Innovationsträgheit der Planwirtschaft in der DDR – Ursachen und Folgen'. *Deutschlandarchiv* 26, pp. 806–18.

Marcus, George (1992) 'Past, Present and Emergent Identities: Requirements for Ethnographies of late Twentieth Century Modernity Worldwide'. In: Scott Lash and Jonathan Friedman (eds.) *Modernity and Identity.* Oxford: Blackwell.

Marx, Karl and Friedrich Engels (1960) *Marx und Engels Werke,* vol. 8. Berlin: Dietz Verlag.

——— (1969) The Eighteenth Brumaire of Louis Bonaparte. In Karl Marx and Friedrich Engels, *Selected Works.* New York, 1969.

Marz, Lutz (1991) 'Der prämoderne Übergangsmanager. Die Ohnmacht des "real-sozialistischen Wirtschaftskaders"'. In: Rainer Deppe; Helmut Dubiel and Ulrich Rödel (eds.) *Demokratischer Umbruch in Osteuropa.* Frankfurt: Suhrkamp.

——— (1992) 'Geständnisse und Erkenntnisse – Zum Quellenproblem empirischer Transformationsforschung'. In: Martin Heidenreich (ed.) *Krisen, Kader, Kombinate,* Berlin: ed. Sigma Bohn.

Mauss, Marcel (1968) *Die Gabe.* Frankfurt: Suhrkamp.

Méda, Dominique (1995) *Le travail. Une valeur en disparition*. Paris: Flammarion.

Merkens, Hans and Dagmar Bergs-Winkels (1990) *Folgen aus Plan und Bürokratie. Die Organisationskultur des VEB Berlin-Chemie*. Arbeitspapier, Berlin: FU, FB Erziehungswissenschaften.

Meuschel, Sigrid (1992) *Legitimation und Parteiherrschaft: Zum Paradox von Stabilität und Revolution in der DDR 1945–1989*. Frankfurt: Suhrkamp.

Morgan, Gareth (1986) *Images of Organisations*. London: Sage.

Müller, Birgit (1992) 'De la "folie du marché" à la fin des illusions'. *Liber* 10, (June).

———(1992) 'De la compétition socialiste à la libre concurrence: la privatisation de trois entreprises (VEB) de Berlin-Est'. *Allemagne aujourd'hui* 121 (July – September).

———(1993) 'Le mur dans la tete. Les stéréotypes interallemands et les problèmes de transition dans trois anciennes entreprises du peuple à Berlin Est'. In: *Les temps modernes*, année 49(560).

———(1993) 'Vom Kollektivmitglied zum "neuen Menschen" in der Markt-twirtschaft: Weltsichten und wirtschaftliches Handeln in drei ehemals volkseigenen Betrieben in Ostberlin'. In: Sabine Helmers (ed.) *Ethnologie der Arbeitswelt*. Mundus Reihe Ethnologie, Band 67. Bonn: Holos Verlag.

Müller, Birgit, Elena Mechtcherkina, Kirill Levinson, Isabelle Cribier and Ondrej Onikienko (1996) '"Wir, die wir einen Floh beschlagen haben …" Produktivität und Profit in einem Moskauer Joint Venture'. In: Birgit Müller (ed.) *A la recherche des certitudes perdues... Anthropologie du travail et des affaires dans une Europe en mutation*. Berlin: Cahiers du Centre Marc Bloch.

Neckel, Sighard (1992) 'Das lokale Staatsorgan'. *Zeitschrift für Soziologie* 21(4).

———(1993) *Die Macht der Unterscheidung*, Frankfurt: Fischer Taschenbuch.

Neunberger, Oswald and Ain Kompa (1987) *Wir die Firma. Der Kult um die Unternehmenskultur*. Basel: Beltz.

Niethammer, Lutz (1990) 'Das Volk der DDR und die Revolution', In: Charles Schmüddekopf (ed.) *'Wir sind das Volk!' Flugschriften, Aufrufe und Texte einer deutschen Revolution*. Reinbek: Rowohlt.

———, Alexander von Plato and Dorothee Wierling (1991) *Die volkseigene Erfahrung. Eine Archäologie des Lebens in der Industrieprovinz der DDR*. Berlin: Rowohlt.

Offe, Claus (1994) *Der Tunnel am Ende des Lichts*. Frankfurt: Campus.

Ouchi, W. G. (1981) *Theory Z*. Reading, MA: Addison-Wesley.

Peters, Tom and Robert H. Waterman (1982) *In Search of Excellence*. New York: Harper and Row.

Polanyi, Karl (1990) *The Great Transformation*. Frankfurt: Suhrkamp.

Preston, Peter (1992) 'Modes of Economic-Theoretical Engagement'. In: Roy Dilley (ed.) *Contesting Markets. Analysis of Ideology, Discourse and Practice.* Edinburgh: Edinburgh University Press.

Rabinow, Paul (1986) 'Representations Are Social Facts: Modernity and Post-Modernity in Anthropology'. In: James Clifford and George Marcus (eds.) *Writing Culture: The Poetics and Politics of Ethnography.* Berkeley: California University Press.

Rifkin, Jeremy (1997) *La Fin du Travail.* Paris: La Découverte.

Rosenlöcher, Thomas (1997) *Ostgezeter.* Frankfurt: Suhrkamp.

Rosenthal, Paul-André (1996) 'Construire le "macro" par le "micro": Frederic Barth et la microstoria'. In: Jacques Revel *Jeux d'echelles.* Paris: Seuil/Gallimard.

Rottenburg, Richard (1991) 'Der Sozialismus braucht den ganzen Menschen. Zum Verhältnis vertraglicher und nichtvertraglicher Beziehungen in einem VEB'. *Zeitschrift für Soziologie* 20(4) pp. 305–22.

———(1992) 'Welches Licht wirft die volkseigene Erfahrung der Werktätigen auf westliche Unternehmen? Erste Überlegungen zur Strukturierung eines Problemfeldes'. In: Martin Heidenreich (ed.) *Krisen, Kader, Kombinate.* Berlin: Sigma Bohn.

Sabel, Charles F. (1982) *Work and Politics.* Cambridge: Cambridge University Press.

Sapir, Jacques (1992) *Logik der sowjetischen Ökonomie oder die permanente Kriegswirtschaft.* Münster: Lit Verlag.

Schlegelmilch, Cordia (1998) 'For the People or with the People? The difficulties of Communal Democracy in a Small Town in Saxony'. In: Birgit Müller (ed.) *Power and Institutional Change in Post-Communist Eastern Europe.* Canterbury: CSA.

Schließer, Waldfried (1990) 'Marktwirtschaft und Sozialismus'. In: Michael Heine (ed.) *Die Zukunft der DDR-Wirtschaft.* Hamburg.

Schmidt, Rudi and Burkart Lutz (ed.) (1995) *Chancen und Risiken der industriellen Restrukturierung in Ostdeutschland.* Berlin: Akademie Verlag.

Scott, James (1990) *Domination and the Arts of Resistance: Hidden Transcripts.* New Haven and London: Yale University Press.

Segert, Astrid (1995) 'Das Traditionelle Arbeitermilieu in Brandenburg. Systematische Prägung und regionale Spezifika'. In: Michael Vester, Michael Hofmann and Irene Zierke, (eds.) *Soziale Milieus in Ostdeutschland.* Köln: Bund-Verlag.

Senghaas-Knobloch, Eva (1992) 'Notgemeinschaft und Improvisationsgeschick: Zwei Tugenden im Transformationsprozeß'. In: Martin Heidenreich (ed.) *Krisen, Kader, Kombinate.* Berlin: Sigma Bohn.

Sofsky, Wolfgang and Rainer Paris (1991) *Figurationen sozialer Macht.* Opladen: Leske und Budrich.

Spülbeck, Susanne (1995) 'Das Jammern und die Besserwessis'. In: Wolfgang Benz and Marion Neiss, *Deutsche Erfahrungen – deutsche Zustände*. Berlin: Metropol Verlag.

Stark, David and Lásló Bruszt (1998) *Postsocialist Pathways. Transforming Politics and Property in East Central Europe*. Cambridge: Cambridge University Press.

Stolcke, Verena (1995) 'Talking Culture'. *Current Anthropology* 36(1).

Sztompka, Piotr (1991) Society in Action. *The Theory of Social Becoming*. Chicago: The University of Chicago Press.

———(1993) *The Sociology of Social Change*. Oxford: Blackwell.

Tambiah, S.J. (1970) *Buddhism and the Spirit Cults in South-East Thailand*. London: Cambridge University Press.

Thompson, Edward P. (1967) 'Time, Work-Discipline and Industrial Capitalism'. *Past and Present* 38.

Ulrich, Bernd (1997) 'Doppeltes Schnäppchen. Globalisierung oder die Angst vor der Diktatur des Jetzt'. *Frankfurter Allgemeine Zeitung* 57 (8 March).

Verdery, Katherine (1996) *What was Socialism and What Comes Next?* Princeton. Princeton University Press.

——— (1998) 'Property and Power in Transylvania's Decollectivization'. In: Chris M. Hann, (ed.) *Property Relations. Renewing the Anthropological Tradition*. Cambridge: Cambridge University Press.

Vester, Michael (1995) 'Milieuwandel und regionaler Strukturwandel in Ostdeutschland'. In: Michael Verster, Michael Hofmann and Irene Zierke (eds.) *Soziale Milieus in Ostdeutschland*. Köln: Bund-Verlag.

Voskamp, Ulrich and Volker Wittke (1991) 'Aus Modernisierungsblockaden werden Abwärtsspiralen – Zur Reorganisation von Betrieben and Kombinaten in der ehemaligen DDR'. *Berliner Journal für Soziologie* 1, pp.17–39.

Wachtel, Nathan (1971) *La vision des vaincus*. Paris: Gallimard.

Wachtel, Nathan (1974) 'L'acculturation'. In: Jacques Le Goff and Pierre Nora (eds.) *Faire de l'histoire*. Paris: Gallimard.

Wagener, Hans Jürgen (1996) 'Transformation als historisches Phänomen'. *Arbeitsberichte des Frankfurter Instituts für Transformationsforschung* 7(96).

Weber, Max (1972) Wirtschaft und Gesellschaft. Grundriss der verstehenden Soziologie, 5 edition. Tübingen: J.C.B. Mohr.

Index

www.ingramcontent.com/pod-product-compliance
Lightning Source LLC
Chambersburg PA
CBHW060033030426

42334CB00019B/2304